Newton's *Sensorium*: Anatomy of a Concept

Archimedes

NEW STUDIES IN THE HISTORY AND PHILOSOPHY OF SCIENCE AND TECHNOLOGY

VOLUME 53

Archimedes has three fundamental goals; to further the integration of the histories of science and technology with one another: to investigate the technical, social and practical histories of specific developments in science and technology; and finally, where possible and desirable, to bring the histories of science and technology into closer contact with the philosophy of science. To these ends, each volume will have its own theme and title and will be planned by one or more members of the Advisory Board in consultation with the editor. Although the volumes have specific themes, the series itself will not be limited to one or even to a few particular areas. Its subjects include any of the sciences, ranging from biology through physics, all aspects of technology, broadly construed, as well as historically-engaged philosophy of science or technology. Taken as a whole, *Archimedes* will be of interest to historians, philosophers, and scientists, as well as to those in business and industry who seek to understand how science and industry have come to be so strongly linked.

More information about this series at http://www.springer.com/series/5644

Jamie C. Kassler

Newton's *Sensorium*: Anatomy of a Concept

Springer

Jamie C. Kassler
Northbridge, NSW, Australia

ISSN 1385-0180 ISSN 2215-0064 (electronic)
Archimedes
ISBN 978-3-030-10152-7 ISBN 978-3-319-72053-1 (eBook)
https://doi.org/10.1007/978-3-319-72053-1

Printed on acid-free paper

This Springer imprint is published by the registered company Springer International Publishing AG part of Springer Nature.
The registered company address is: Gewerbestrasse 11, 6330 Cham, Switzerland

Introduction

In the writings that Newton intended for a public audience, he used the term *sensorium* in relation to both God and humans. The first usage has attracted the notice of a number of commentators, whereas the second usage has had little serious attention, so that, of course, there has been no systematic investigation into the relationship between the two usages. As a consequence, the divine *sensorium* has been glossed to mean the world, the mind of God, absolute space and, sometimes also, absolute time.[1] But the human *sensorium* has been glossed more narrowly to mean either the nervous system, the brain or a particular, but unidentified, part of the brain.[2] Now, the literal meaning of the Latin term *sensorium*, or its English equivalent 'sensory', is 'thing that feels'. But this meaning offers little insight into Newton's concept, because it is a construct, that is to say, a concept specially devised for a theory.

In the following inquiry, I attempt to elucidate the meaning of his concept by discovering its underlying model, beginning in Parts II and III with the *sensorium* in relation to humans. This part of his concept is situated, first, in the context of his own writings and, then, in the context of certain seventeenth-century developments in anatomy and physiology. Only then is it possible to draw conclusions about the *sensorium* in relation to God. For, as will be evident in Part IV, what Newton called 'the analogy of nature' is a shorthand term for his method of reasoning from experience; and this method requires that the second part of his concept is last in the order of knowledge, as will become evident towards the end of this inquiry.[3]

---

[1] See, e.g., Burtt, *The Metaphysical Foundations of Modern Physical Science*, p. 261: 'Absolute space is the divine sensorium'. Hall, 'Henry More', XIII, p. 49: 'God ... with the world as his sensorium'. Henry, 'Voluntarist Theology', p. 110 n.61: 'phantasms in the mind (or "sensorium") of God'. Holton, 'Presupposition in the Construction of Theories', p. 92: 'space and time ... called the "sensory" of God'. Yolton, *Thinking Matter*, p. x: 'space is ... the sensorium of God'.

[2] See, e.g., Brook, 'Beyond Everything', p. 24: 'sensorium (nervous system)'; Burtt, *The Metaphysical Foundations of Modern Physical Science*, p. 235: 'a particular part of the brain' called 'the *sensorium*'. Shapiro, *Fits, Passions, and Paroxysms*, p. 73: 'the brain, or sensorium'. Yolton, *Thinking Matter*, p. x: 'our brain is the sensorium of our ideas'.

[3] More recently, Newton's term 'analogy of nature' has been substituted by the term 'transdiction'

I call my study 'an inquiry', because it involves a process of discovery, as well as an exploration of problems. Consequently, the summary below of each of its four parts is intended merely as a guide for the reader, rather than a revelation of what will be discovered along the way. Nevertheless (and by way of anticipation), I shall mention one discovery here, namely, that Newton's construct involves a spectator, as well as a spectacle. But this should not be surprising, since as long ago as 1950 Henry Margenau pointed out that 'the spectatorial doctrine has gone hand in hand with the rise of one special branch of physics, [namely,] mechanics, and is indeed its logical correlate'.[4] In making this statement, however, Margenau imagined the spectator as a physicist and the spectacle as masses moving within a stationary 'container' of Newtonian absolute space. But in the course of this inquiry, the spectator will emerge as a principle of intellectual life, the hieroglyph for which is an intellectual eye. In the case of humans, the intellectual eye views the external spectacle indirectly as internal representations in the space of its *sensorium*, whereas in the case of God, the intellectual eye beholds directly the cosmic spectacle situated within a finite region of space, the whole of which is the space of its *sensorium*.

The inquiry proper begins in Part I with twelve texts extracted from Newton's writings, because they either mention or provide a context for his term *sensorium* in relation to both humans and God. Although this term is found in some of Newton's personal papers, during his lifetime access to his papers was restricted to only a few of his inner circle, and after his death most of his papers remained unpublished until scholarly interest in them gradually increased after World War II.[5] Accordingly, the texts presented here have been extracted from sources intended for a public audience, since these sources laid the foundation for the reception of Newton's concepts.

Although the twelve texts themselves are provided with brief notes and comments, those related to the human *sensorium* are singled out for analysis in Part II, beginning with a text in which there is no mention of the term *sensorium*. But this text is important, not only because it presents a potential model for Newton's theoretical construct but also because it provides insight into the scope of Newton's concept in relation to the sensory-motor system. From an analysis of this and other texts, some general conclusions are drawn, followed by a summary of what has been discovered thus far. But the summary also points to four questions that cannot be answered either from Newton's texts or from an analysis of them.

(see Mandelbaum, *Philosophy, Science and Sense Perception*, pp. 61–117) and afterwards by the now preferred term 'transduction' (see Shapiro, *Fits, Passions, and Paroxysms*, pp. 4–5, 40, *et passim*).

[4] Margenau, *The Nature of Physical Reality*, p. 35. This important book on the epistemological problems of 'classical' (i.e. Newtonian) and modern mechanics (i.e. statistical, quantum, relativity) also includes three chapters devoted to the problem of passing from sensory awareness to orderly knowledge, the departure from the immediate (constructs) and the metaphysical requirements on constructs.

[5] For a brief description of the current location of these papers, see Westfall, *Never at Rest*, pp. 875–7; and for a brief description of the current efforts to make the papers more accessible, see Guicciardini, 'Digitizing Isaac Newton'.

Consequently, Part III seeks answers to the four questions by exploring the work of two English physicians—Thomas Willis (1621–75) and William Briggs (c.1650–1704). But I treat the younger physician first, partly because he was a personal acquaintance of Newton and partly because of his special study, ophthalmology, the science of the anatomy, physiology, diseases and treatment of the eye. By contrast, Willis's special study, the science of 'neurologie', was far more extensive, since it included the entire apparatus of the human body in health and disease. Indeed, a modern commentator has described his 1664 book, *Cerebri anatome*, as 'the foundation document of the anatomy of the central and autonomic nervous systems'.[6]

It might be asked why I have chosen the work of Willis and Briggs as a key to Newton's concept, although neither physician is mentioned by name in the writings he made public. The short answer is that the two physicians, who were at the forefront of their respective specialties, helped to familiarise the English medical profession with the 'correct' theory of physiological optics, according to which the retina, as an expansion of the optic nerve, is the receptor of images of external objects.[7] The foundation of that theory was laid in 1583, when the physician, Felix Platter, identified the retina as the receptive part of the eye; but the fully developed physiological theory was not produced until some 20 years later by Johannes Kepler, who thus contributed the 'first' important knowledge of visual space perception.[8] Although his new theory was adopted by philosophers such as René Descartes, Robert Hooke, Christiaan Huygens and Newton himself, nevertheless there was continuing dispute as to whether the retina or some other part of the eye was the receptor for sight. Indeed, Briggs's writings reveal that the dispute had not been settled even by the 1680s.[9]

The new physiological theory of *optical* processes did not explain how the inverted image formed on the retina was perceived by the mind or, in modern terms, the brain. Nor could the new theory explain this, because perception involves *visual* processes, the understanding of which was in its infancy. Although it is known today that the brain infers, interprets, evaluates and compares, yet it is not understood, for example, how it interprets colour which is not present in what reaches and enters the eye. Nevertheless, Willis and Briggs believed that visual sensation occurs only when the stimulation from the retina reaches a part *in* the brain which they termed '*sensorium commune*' or 'common sensory'. But Willis alone provided a detailed anatomical description of the common sensory, identified its specific place in the anatomical space of the brain and provided a new theory about the 'office' and use of this particular entity.

---

[6] Frank, Jr., 'Thomas Willis', p. 406, who also indicated that, until the late eighteenth century, the work in question was used as a text and, until the mid-nineteenth century, it was mandatory background reading for neuro-anatomists

[7] See Koelbing, 'Ocular Physiology', p. 233, who referred only to Briggs, since he was unaware of Willis's contribution.

[8] See Boring, *Sensation and Perception*, p. 97.

[9] See *infra* Pt.III.3.1, p. 70 n.59.

The exploration in Part III of the concepts of Briggs and Willis does not provide certain answers to the four questions raised at the conclusion of Part II, but it does provide insight into how those questions might be answered. Since they also relate to the divine *sensorium*, I reconsider the questions at the outset of Part IV in order to provide plausible answers. I then turn to Newton's texts extracted from his teleological arguments, for these provide data for assessing his variety of theism and his functional conception of God as an 'intelligent Agent'. Finally, after a brief exposition of Newton's method of reasoning from experience, I address and attempt to solve the problem of the divine *sensorium* and, in the concluding section of Part IV, to assay a related problem concerning whether or not Newton's absolute space is a container, as many commentators have supposed.

During the discovery process that follows, it is important to keep in mind that the seventeenth century was a period of transition when classical theories, as well as newly introduced theories, formed the background for discussions and disputes in anatomy and physiology.[10] Indeed, because the 'new' mechanical philosophy of Descartes spread rapidly at Cambridge University during the 1640s, his ideas also became a major catalyst for discussion and debate among university members, some of whom were alarmed at their implication.[11] Even as late as 1667, the year Roger North was admitted to the same university, he reported hearing 'a sort of sly discours' about this new philosophy, which was regarded as 'a sort of heresie'.[12] Note, therefore, that in his seminal study of Newton's mathematical methods, Niccolò Guicciardini has shown that, between the 1660s and 1680s, Newton began 'elaborating a deep distaste for all things Cartesian, and, at the same time, developing a veneration for certain ancient [mathematical] traditions'.[13]

Certainly, Newton did not completely escape the influence of Descartes, and the same may be said of many others, including Willis. For it was Descartes's commitment to a total mechanistic hypothesis of nature that enabled him to revolutionise the problem of how the perceiving organism is related to the world perceived. Nevertheless, both Willis and Newton believed that the French philosopher had carried mechanism too far. As a consequence, the former attempted to re-vitalise the 'machine' of the human body and the latter the 'machine' of the cosmos.[14] And, as will be apparent later on, they did this by resorting to incorporeal as well as corporeal disposing principles, powers or properties conceived as unknown causes that produce effects.

[10] For instances of this, see Frank, Jr., *Harvey and the Oxford Physiologists*.

[11] See Feingold, 'Isaac Barrow', p. 29.

[12] See Kassler, *Seeking Truth*, pp. 19–21, and sources cited there.

[13] The tradition in question derived chiefly from the founder of the Alexandrian school, Euclid; see Guicciardini, *Isaac Newton*, p. 59, *et passim*. However, as Guicciardini also noted, p. 368, Newton's mathematical methods are 'a Cartesian heritage', so that, p. 104: 'There is often something illogical in Newton's anti-Cartesian invectives'.

[14] This concern was addressed in quite different ways, e.g., by the theologian, Henry More, and by the physician, Francis Glisson. For the former's spiritualist vitalism, see *infra* Pt.II.2.2, pp. 53–4; for the latter's hylozoic vitalism, see Temkin, 'Francis Glisson'. See also Boyle, *Free Enquiry*, for these and other vitalisms.

During the course of the following inquiry, which is evidence-based, not speculative, a number of discoveries will be made that are incommensurate with certain claims about Newton in the secondary literature. Since these claims cannot be addressed within the constraints of an inquiry, I shall mention here and comment briefly on claims that concern four topics (a fifth is dealt with in the concluding section of Part IV). The topics are (1) the traditions that influenced Newton, (2) his theology, (3) his metaphysics and (4) his notion of a human *sensorium* in the context of mind-body relations. Absent from this list is the topic of the divine *sensorium*, because the secondary literature on this topic completely ignores the importance of Newton's method of reasoning from experience, which proceeds from the known to the unknown or, more accurately, from that which has been observed to be universally true to that which is beyond the limits of observation.[15]

(1) Many commentators have claimed that Newton's thought derives from one or another tradition, including but not restricted to atomism, Platonism in its various guises, Pythagoreanism, Stoicism and Cartesianism. But this topic also presents a general problem that recurs in the other topics discussed below, namely, the problem of what constitutes a tradition, a problem that may be captured in the form of two questions: First, did Newton consistently follow one or another philosophical tradition? Or, second, did he merely absorb a plethora of influences which he then adapted to his own purposes?

An affirmative answer to the first question was given in 1966 by J. E. McGuire and P. M. Rattansi, who interpreted Newton's comments on the wisdom of the ancients as references to a monolithic tradition that supposedly dates back to a 'pristine age' before the corruption of Noah and his sons.[16] Since these commentators claimed that the tradition in question was still alive in Newton's day, of course they were able to point to a multitude of persons, who over time might have influenced his thought.

In 1984 the McGuire and Rattansi thesis was subjected to critical examination in an important introduction to Newton's so-called Classical Scholia.[17] Its author, Paolo Cassini, then put forward a different thesis, according to which Newton's own doxography of the wisdom of the ancients may be interpreted as the use by him of 'a non-dogmatic principle of [textual] authority, turned to the legitimization of action at a distance within the framework of the new mathematical astronomy'.[18] Thus:

> Among the ancient philosophers—atomists, pre-Socratics, Platonists, Pythagoreans—Newton did not so much seek for a broad revelation as, particularly, for the fundamental technical propositions of his own mathematical astronomy; more precisely, he 'read' into their testimony the law of gravitational attraction. Obviously in so doing he had no predecessors.[19]

---

[15] See Jevons, *The Principles of Science*, p.vii.

[16] See McGuire and Rattansi, 'Newton and the Pipes of Pan'.

[17] Cassini, 'Newton: The Classical Scholia', pp. 3–7, 13–4

[18] Ibid., p. 6

[19] Ibid.

Cassini also provided evidence that the same principle of textual authority had been used earlier by 'protagonists of the astronomical revolution' such as Copernicus, Galileo, Kepler and their followers, who 'often understood the progress of astronomy as being a reversion towards propositions comprehended intuitively by the Ancients'.[20] Note, therefore, that this understanding of progress was not restricted to astronomers, for in the prefatory pages of *De humani corporis fabrica* (1543), the anatomist, Andreas Vesalius, also used a strategy similar to the one Cassini ascribed to Newton.[21]

Prior to Cassini's critique, Betty Jo Dobbs supported the McGuire and Rattansi thesis in her extensive 1975 investigations into Newton's unpublished alchemical writings,[22] whereas by 1985 she concluded that Newton was an 'eclectic thinker'.[23] In changing her mind, she seemed to give an affirmative answer to the second question raised above. Ironically, however, Dobbs's answer could also be applied to the McGuire and Rattansi thesis, because historical circumstances would lead to many different variants of the supposed monolithic tradition. And this in turn suggests that the claim of such a tradition is unsustainable.

At the core the McGuire and Rattansi thesis is the idea of corruption from a pristine or former state, so that it might be argued that the recurrence of such an idea is a pattern of thought, not a tradition. Indeed, this pattern is found in the early chapters of Genesis concerning the expulsion of Adam and Eve from the paradise of the earthly Garden of Eden, because they had eaten the forbidden fruit of the tree of life (later symbolised as the tree of knowledge of good and evil). According to one commentator, this paradise myth has served as a 'primal Utopia',[24] so that its pattern recurs not only in later prophetic writing and classical poetry but also in many subsequent utopian writings that exhibit a desire for a return to, or restoration of, a golden age.[25]

For example, this pattern recurs in the Middle Ages, when the fourth-century church father, Eusebius of Caesarea, first proposed the myth that a morally pure

---

[20] Ibid., p. 10

[21] According to Henry, 'Galen's Mistakes', Vesalius provided not only 'a defiant apologia for deviating from the received anatomical wisdom of Galen' but also 'a custom-made history of medicine, in which he himself appeared as the reviver of the ancient Greek practice of performing dissections on humans'.

[22] Dobbs, *The Foundations of Newton's Alchemy*, pp. 235 *seq*.

[23] Dobbs, 'Newton and Stoicism', p. 111. Following in the footsteps of Dobbs, a recent study attempted to determine whether Newton's eclecticism might have a metaphysical basis due to 'Neostoic and Neoplatonic influences'; see De Smet and Verelst, 'Newton's Scholium Generale'. For a critical comment on this attempt, see Ducheyne, 'General Scholium', p. 3 and n.18.

[24] For the context, see the entire chapter on 'The Messianic Kingdom' in Ferguson, *Utopias of the Classical World*, pp. 146–53.

[25] For Newton's own treatment of myth in his religious and chronological studies, see Buchwald and Feingold, *Newton and the Origin of Civilization*, pp. 141–63. According to the evidence provided there, Newton had recourse to the idea of a golden and other ages in *Works and Days* of Hesiod. According to Kirk, Raven and Schofield, *The Presocratic Philosophers*, p. 34, this ancient author represents a 'quite new attempt to systematize the ancient myths'.

primitive Christian church was corrupted over time. And this myth was later promulgated by one of Newton's acolytes, Samuel Clarke, who sought to restore the supposed purity of Christianity as practised in the second and third centuries, when the church was Arian.[26] But, of course, a recurring pattern of thought neither solves the problem of how the term 'tradition' is to be understood nor the problem of what methodologies might be required for assigning Newton to one or another tradition.

(2) Since the early part of the twentieth century, there has been a growing consensus among commentators that Newton's theology belongs to the tradition called 'voluntarism', a term that applies to any philosophical theory according to which the will is prior or superior to the intellect. But the term itself was introduced into the philosophical literature towards the end of the nineteenth century, after which there has been an increasing proliferation of classes of voluntarist theories, including epistemological, ethical, legal, metaphysical, psychological and theological. What is more, subclasses of some of these main classes have also been identified. Insofar as Newton is concerned, commentators have classed him as a theological voluntarist, probably following Edwin Burtt, who in 1924 asserted that 'in common with the whole voluntaristic British tradition in medieval and modern philosophy', Newton 'tended to subordinate in God the intellect to the will' and, consequently, he stressed 'God's power and dominion' above 'the Creator's wisdom and knowledge'.[27]

Since this kind of interpretation of Newton's theology continues to the present time, Peter Harrison, in a paper published in 1990, dared to question the description of Newton as a voluntarist. Unfortunately, he hedged that question by suggesting that there was a 'more important' one, namely, 'whether voluntarism is a helpful way of distinguishing theological motivations and natural philosophical commitments of this [early modern] period'.[28] In a subsequent paper, published in 2002, he took up this supposed 'more important' question not only in relation to the term's theological meaning, but also in relation to the 1961 thesis of Francis Oakley that certain developments in medieval theology mark 'the beginning of that fruitful stream of *voluntarist* natural law thinking' in science.[29] Obviously, this strategy cannot answer the question that Harrison raised initially and failed to answer, perhaps because he implicitly assumed that the central problem is the term 'voluntarism', not the term 'will'.

In a paper published in 2009, John Henry responded to Harrison's second paper in order to 're-affirm, contrary to Harrison's claims, that voluntarist theology was an important component, or at least concomitant, of the natural philosophy of some of

---

[26] See Kassler, *Seeking Truth*, pp. 214–5. For an instance in which Newton himself resorts to the pattern of corruption from a pristine state, see p. xxi n.85 below.

[27] See Burtt, *The Metaphysical Foundations of Modern Science*, p. 294.

[28] Harrison, 'Was Newton a Voluntarist?', p. 61

[29] See Harrison, 'Voluntarism and Early Modern Science', pp. 63 and 79; see also Oakley, 'Christian Theology and the Newtonian Science'.

the leading thinkers of the early modern period'.[30] Of course, he included Newton as one of the leading thinkers. But in the few pages devoted to him, Henry directed much of his attention to an earlier paper by Harrison on miracles.[31] Nevertheless, as support for Newton's voluntarism, he quoted a snippet of data consisting of an incomplete sentence taken from Query 31 of the modernised Dover version of the 1730 edition of Newton's *Opticks*: that God 'is more able by his Will to move the Bodies within his boundless uniform Sensorium, and therefore to form and reform the Parts of the Universe, than we are by our Will to move our own Bodies'.[32] There are several points to note about this snippet.

First, the snippet is extracted from a longer argument from design, in which Newton points to uniformities in both cosmic bodies and bodies of animals as evidence that such uniformity can only be 'the effect of nothing else than the Wisdom and Skill of a powerful ever-living Agent who being in all Places, is more able by his Will to move the Bodies within his boundless uniform Sensorium and therefore', etc.

Second, Query 31 is one of only two texts in which Newton uses *terms for God as* a causal 'Agent' in the world. For in the other texts in Part I extracted from his arguments from design, Newton uses *terms for God as* 'an incorporeal, living, intelligent and ever-present being', *as* 'the intelligence and wisdom of a powerful ever-living being' and *as* 'a Being incorporeal, living, intelligent'. The chief import of the term 'intelligent' is that the 'Being' in question is rational and, therefore, God possesses an intellect.

Third, Query 31 and all the other passages occur within the context of one or another of Newton's arguments from design, a context that places constraints on interpretations of the passages, because we cannot know the design in the mind of God, only its effects. But Henry, who made no mention of this context, also ignored the other passages quoted above in which stress is laid on God as a rational being, no doubt because these passages would make it difficult for him to sustain his belief that Newton's theology is voluntarist, that is to say, that, for him, will is superior to intellect.

In a paper published in 2017, Henry and his co-author, J. E. McGuire, again sought to retain Newton's supposed theological voluntarism by turning this theology into panentheism. According to them, 'this theology maintains that God and the world are ontologically distinct, but [that] the world exists in God; and the inclusion of the world in God is held not to be the totality of God, which is more than the world'.[33] But does panentheism claim that the will is superior to the intellect? The authors did not address this question, nor did they make a coherent or evidence-based argument that panentheism is a synonym for voluntarism.

---

[30] Henry, 'Voluntarist Theology at the Origins of Modern Science: A Response to Peter Harrison', p. 79

[31] Harrison, 'Newtonian Science, Miracles, and the Laws of Nature'

[32] Ibid., p. 92

[33] Ibid., p. 14

Although they claim that the term 'panentheism' was coined in 1829, they do not indicate that the etymon is German ('Panentheismus'), that the term itself is absent from the main as well as the supplementary volumes of the multivolume *Oxford English Dictionary* or that the more recent online version of that dictionary indicates that its first English usage was in 1874. Perhaps to avoid this problem, the authors claimed that panentheism can be traced back 'at least to Plotinus' and that 'Neoplatonic panentheism' enjoyed 'something of a revival during the Renaissance and Reformation'.[34] But is the supposed tradition of panentheism the same as that of voluntarism?

Like the previous papers commented on above, Henry and McGuire failed to provide a positive account of what Newton means by the term 'will'. Surely, such an account is a basic requirement for determining whether or not Newton is a voluntarist (of whatever stripe). It is therefore worth pointing out that as long ago as 1964 Vernon Bourke addressed the question, 'what does will mean?' (there is not one meaning but several); demonstrated the difficulties in using the label 'voluntarism'; provided a definition of the general theory of legislative will; treated some special theories, including legal voluntarism and legal intellectualism; and pointed out that, as a young writer, John Locke was 'a thorough legal voluntarist' but afterwards wrote of the law of nature as the law of reason.[35]

(3) Just as there have been numerous papers arguing for Newton as a voluntarist, so too there have been numerous papers attempting to assign some kind of metaphysics to Newton.[36] Indeed, as one commentator complained in 1993: 'The majority of historians of ideas have been reluctant to banish metaphysical considerations from the mind of the author of the *Principia*'.[37] And I would add that in many of the papers on Newton's metaphysics, the problem of tradition recurs when attempts are made to assign a particular kind of metaphysical tradition as the supposed source for Newton's metaphysics

One exception to this rule may be found in a book-length treatment of the topic published in 2008, in which no claims are made for philosophical traditions as the metaphysical basis of Newton's thought. Instead, the main thesis of its author, Andrew Janiak, is that Newton 'consistently articulated a compelling and overarching conception of the relation between mathematical physics on the one hand, and more clearly metaphysical concerns on the other'.[38]

To develop his thesis, Janiak presented an interpretation of Newton's natural philosophy as bifurcated into a divine and a mundane metaphysics. Divine

---

[34] Ibid., pp. 14–5

[35] See Burke, *Will in Western Thought*.

[36] See, e.g., Zafiropulo and Monod, *Sensorium Dei*, who claimed that the divine *sensorium* invoked by Newton had its roots in the older metaphysics of Boethius. For a different suggestion, see p. x n.23 above.

[37] See Hall, 'Henry More and the Scientific Revolution', XIII, p. 50. For one answer to such historians of ideas, see Levitin, 'Newton and Scholastic Philosophy'.

[38] Janiak, *Newton as Philosopher*, p. 7.

metaphysics, which is fixed and not subject to revision, represents Newton's 'fundamental conception of God's nature and relation to the natural world', whereas mundane metaphysics, which 'occurs within the basic framework centered on the divine', is subject to 'precisely the sorts of revision and refinement that characterize all of Newton's other work'.[39] Newton, therefore, endowed God with a 'metaphysical, but not an epistemic, primacy'.[40]

For his interpretation, Janiak relied on two of Newton's writings: the first, on which his interpretation was based, is the posthumously published 'De gravitatione'; the second, to which his interpretation was extended and developed, is the 1687 edition of *Principia mathematica*. But, in fact, the second 1713 edition is cited much more frequently, because, according to Janiak's exaggerated claim, Newton discusses God 'extensively' in the 'General Scholium', which was added to this edition.[41] Unfortunately, in arguing his case for Newton's supposed metaphysics, Janiak relied more on his philosophical colleagues, as recorded in the book, rather than on a careful study of 'De gravitatione', the manuscript on which his interpretation was based.

This manuscript was a product of Newton's mathematical lectures at Cambridge University read during the 1670s to an audience of students; and its title is taken from the first incipit of the manuscript, 'De gravitatione et aequipondio fluidorum et solidorum in fluidis', which suggests that the subject of the lecture is hydrostatics.[42] Although some pages at the conclusion are devoted to this subject, the main body of the manuscript contains arguments based on mathematical hypotheses, definitions and concepts.

For example, Newton states that he does not define 'body' (corporeal substance) in 'a philosophical manner', so that 'instead of physical bodies', the students are to understand 'abstract figures in the same way that they are considered by Geometers when they assign motion to them, as is done in Euclid's *Elements*, Book[s] 1, 4 and 8'.[43] And at the conclusion of this portion of the manuscript, he again reminds his student audience that he has accommodated his definitions 'not to physical things but to mathematical reasoning, after the manner of Geometers'.[44] But Janiak failed to mention the mathematical context of Newton's manuscript.

---

[39] Ibid., p. 45

[40] Ibid., p. 48. According to the author, pp. 8–9, Newton 'never presents an overarching theory of knowledge'; rather, 'the primary epistemic questions' are raised only by 'physical theory'. The term 'overarching' is never defined, although it appears with some frequency in the book. Consequently, if Newton's methodology is understood as 'overarching', then, as I believe, this methodology constitutes his epistemology of science.

[41] See Janiak, *Newton as Philosopher*, p. 151 n.37. Note that the term 'extensively' is more appropriately applied to the arguments from design in certain queries that were added to Newton's 1706 *Optice*. But Janiak did not include these in his interpretation.

[42] See Newton, 'De gravitatione', p. 121.

[43] Ibid., p. 122

[44] Ibid., p. 150

In the same portion of the manuscript, there is a 'digression'[45] relating to Descartes's conception of space as the indefinitely extended impenetrable 'body', a conception that Newton seeks to replace with his own mathematical hypothesis that space and matter are distinct. Accordingly, he attempts to ensure that the concepts arising from this hypothesis are consistent with, as well as confirming, the 'chief truths' of ontology or the science of being.[46] Note, therefore, that Isaac Barrow and Henry More, two of Newton's older contemporaries at Cambridge, not only shared the hypothesis that matter and space are distinct but also explored some of the ontological problems relating to it. But their exploration was guided by their respective interests, Barrow's being mathematical and More's theological, whereas Newton's exploration of these problems is closer to Barrow's[47] than to More's.[48]

Apparently, Janiak was unaware of the forementioned hypothesis, so that he provided a rather peculiar gloss on Barrow and More, both of whom 'contemplate counterfactuals regarding the status of space in a divinely inhabited world with no, or with spatially restricted, matter'.[49] In the case of the theologically inclined More, however, Janiak did acknowledge the latter's claim that space is an 'emanative effect' of God.[50] And since in 'De gravitatione' Newton makes a similar claim,[51] the so-called emanative effect thesis is put to considerable use in Janiak's book, even though the writings Newton published in his own lifetime do not use the phrase 'emanative effect'.

This suggests that More's theology may have been a source for Janiak's interpretation of Newton's 'De gravitatione'. For example, in a 1659 book that Newton read and later owned,[52] More argued that all substances, *corporeal and incorporeal*, are alike in being extended, whereas in Janiak's interpretation no distinction is made between the two 'natures'. Instead, he claimed that just as all substances in the world are extended, so too are God, mind and body—God as an 'infinite penetrable extended substance', mind as a 'finite penetrable extended substance' and body as an 'impenetrable extended substance'.[53] This claim implies that Janiak identified God and mind, respectively, with infinite and finite space and body with the indefinitely extended impenetrable of Cartesian space. What is more, according to him, since the 'spatiotemporally ubiquitous' God is 'actively present' at every place in space and at every moment in time', he therefore 'acts locally, just as any other

[45] Ibid. 123–48, on the last page he states: 'I have already digressed enough'.

[46] Ibid., p, 142

[47] Henry and McGuire, 'Voluntarism and Panentheism', p. 12

[48] For More's early and mature conceptions of space, see Mackinnon in More, *Philosophical Writings*, pp. 283, 293–5; see also the abstract of his *Enchiridion metaphysicum, sive de regus in corporeis dissertatio* (1671), in *Philosophical Transactions of the Royal Society* (1671), 6: 2182–4.

[49] Janiak, *Newton as Philosopher*, p. 142 n.20

[50] Ibid., pp. 243–4

[51] See Newton, 'De gravitatione', pp. 132, 136, 137.

[52] See *infra* Pt.II.2.2, p. 54.

[53] Janiak, *Newton as Philosopher*, p. 177

substance does' and, consequently, *not only God but also the 'objects of metaphysics have become physical'*.[54]

By the time the reader comes to this statement at the end of the book, he or she is apt to forget Janiak's much earlier statement that 'God's relation to the "system of the world" forms a framework within which physical research takes place'.[55] Accordingly, the term 'physical' is one of Janiak's problematic terms, because *Principia mathematica* is a book on rational mechanics and, hence, treats not only the mathematical principles of natural philosophy but also mathematical methods that helped to transform natural philosophy into what we now call 'physics'.[56] Indeed, *Principia mathematica* represents not only an endpoint of a tradition but also a new beginning in mechanics.[57]

As long ago as 1936, Edward Strong critically examined the assumption that 'the edifice of modern science was built on foundations provided by metaphysical architecture',[58] after which he concluded that:

> Modern mathematical-physical science established its method and achieved its results in spite of, rather than because of ... [the] metamathematical tradition.[59] Had the early-modern mathematical investigators in general, rather than by exception, taken the philosophical tradition seriously, history might have seen more mixtures of metaphysics and science similar to Kepler's, without, perhaps, the saving conditions that brought Kepler's metaphysical predispositions to a scientific issue.[60]

In his critique Strong mentioned Newton only in passing. But in a paper published in 1952, he examined the relation between Newton's theism and his natural philosophy by developing a demarcation thesis.

According to his thesis, 'operations assigned to God are expressions of faith and known to be such by Newton', and, hence, any 'principles thereby announced would be theological rather than mathematical or mechanical principles'. For had 'Newton himself failed to differentiate tenets of faith in religion from rules of reasoning in science, there would be an initial likelihood in favor of the argument that theological principles of God's dominion and power were foundational to his natural philosophy'.[61] In 2006 support for this thesis was provided by Steffen Ducheyne,

---

[54] Ibid., p. 165 (italics mine)

[55] Ibid., p. 49

[56] See the *Oxford English Dictionary* for the more common seventeenth-century terms 'natural philosophy', 'natural science' (i.e. knowledge of nature) and 'physiology'. The last term, 'physiology', had two meanings. The first, which is now obsolete but in use between 1564 and 1797, denoted the study and description of natural objects or a particular system of natural science or natural philosophy. The other meaning, in use from 1597, denoted the science of normal functions and phenomena of living beings.

[57] See the concluding chapter on Newton in Dijksterhuis, *The Mechanization of the World Picture*, whose classical study is not cited by Janiak.

[58] Strong, *Procedures and Metaphysics*, p. 4, whose target was Burtt, *The Metaphysical Foundations of Modern Science* (the first edition of which was published in 1924)

[59] That is, Pythagoreanism, as well as Platonism, in the extreme version of Plotinus.

[60] Strong, *Procedures and Metaphysics*, p. 217

[61] Strong, 'Newton and God', p. 149

who quoted the following passages from two of Newton's manuscripts: 'Natural philosophy should be founded not on metaphysical opinions, but on its own principles and [end]'; and religion and philosophy are to be 'preserved distinct', for we 'are not to introduce divine revelations into [natural] Philosophy nor philosophical opinions into religion'.[62]

(4) As indicated at the outset, commentators have tended to gloss but not systematically investigate what Newton might have meant by a human *sensorium*. But two papers, published, respectively, in 2014 and 2017, have actually attempted to offer a brief elucidation of his meaning. In the earlier paper, the principal purpose of its author, Patrick Connolly, was to provide a 'systematic' interpretation of what Newton meant by a divine *sensorium*, which he based on snippets of data from Queries 28 and 31 in the 1718 version of Newton's *Opticks*. Unfortunately, like so many other treatments of this subject, his data consists of an incomplete sentence from each query without any mention of the context in which these snippets of data appear. Nor did he recognise that the divine *sensorium* is mentioned by Newton in other versions of the *Opticks*. It is not surprising, therefore, that for him, 'what *exactly* Newton had in mind is not at all clear'.[63]

As part of the preliminaries to his main purpose, Connolly canvassed such topics as the role of 'Sensoria in early modern philosophy', 'Newton and Descartes on extension' and 'Newton's positive account of mind-body interaction'. In considering the first topic, Connolly began with a two-sentence pre-history on Aristotle followed by a one-sentence statement on the *sensorium* in medieval philosophy. He then turned to the early modern period, when, he claimed, that 'the sensorium had become reified ... [as] a physiological object, a part of the body'.[64]

Accordingly, to get 'a better sense' of how the term was used in the seventeenth century, Connolly provided a two-paragraph account of its usage by the mechanical philosopher, Descartes, and by the natural philosopher, Willis.[65] But Connolly notwithstanding, Descartes did not use the term *sensorium* in the two main books relating to his mechanistic physiology.[66] So that it is not accurate to claim that the French philosopher 'identified the sensorium with the pineal gland';[67] what he did do was to make that gland the anatomical 'seat' of common sense and imagination. Nor is Connolly correct to claim that in Descartes we get 'a much more famous formulation of the view found in Willis'.[68] Indeed, as will become evident in Part III, Willis's 'view' is not the same as that of Descartes.

---

[62] See Ducheyne, 'The General Scholium', pp. 15 and 22 n.127, whose quotes are from Newton's manuscripts in Cambridge University Library, Add MS 3965.9 (early 1710s) and Keynes MS 6 (post-1710).

[63] Connolly, 'Newton and God's Sensorium', p. 186

[64] Ibid., p. 186

[65] Ibid., p. 187

[66] See Descartes, *Discourse on Method* and *Treatise on Man*. Note, however, that he did use the term common sense (*sensus communis*).

[67] Connolly, 'Newton and God's Sensorium', p. 187

[68] Ibid.

As for Connolly's second topic, when noting the differences between Descartes and Newton on the nature of the soul, he claimed without acknowledging Janiak that Newton's soul is extended, like 'all existing things'.[69] This claim leads to the final topic noted above, namely, the role of the human *sensorium* in mind-body interaction. Having consulted two of Newton's early writings—his student notebook[70] and 'De gravitatione'—Connolly's chief claim concerning this new topic is that the *sensorium* is 'the place where the body acted on the soul', as well as 'the place where the soul acted on the body'.[71]

How he could make this claim is puzzling, even though he apparently read but did not understand the portion in the student notebook now cited as 'Of Colours' (*c.*1666). For here Newton explores visual processes and, hence, his concern is the physiology of vision. In these explorations, he uses the term *sensorium* several times, but he does not use the term 'soul' (or as Connolly often calls it 'mind'), whereas in 'De gravitatione' the opposite is true, for there he uses the term 'mind' a few times, but he does not use the term *sensorium.*

As for the authors of the 2017 paper, Henry and McGuire, they have already been referred to under (2) above. Stimulated by Connolly's paper, which they cite with some frequency, they relied on the same snippets of data from Queries 28 and 31, though extracting it from the 1730 edition of Newton's *Opticks* in the modernised version published by Dover. Moreover, like Connolly, they provided no information as to the context in which the snippets are found in the various versions (including the Latin version) of Newton's *Opticks*, even though, in all of those versions, Newton's references to the divine *sensorium* are in the context of his arguments from design.

Realising correctly that Newton is making an analogy between the human and the divine *sensorium*, Henry and McGuire devoted more than Connolly's two paragraphs in their attempt to elucidate what Newton meant by the human *sensorium.*[72] Nevertheless, their interpretation, which has signs of haste, is paper-thin, because their evidence is insufficient and their argument incoherent. As a result, there are two egregious errors in their account. The first error is their claim that the incorporeal soul is identical with what Newton calls a 'sensitive substance' in the human *sensorium.*[73]

But, as will be discovered during the inquiry that follows, Newton's 'sensitive substance' is a receptor of stimuli, not a soul; and this receptor is one of Newton's unobservable and, hence, hypothetical corporeal substances that mediate function, in the case of the sensitive substance, sensory function. Not having read Newton's *Opticks* carefully or even the pertinent literature on Newton's assumptions concerning these hypothetical substances, they opined that the supposed identity between

---

[69] Ibid., p. 189

[70] For the notebook in question, see McGuire and Tamny, *Certain Philosophical Questions*.

[71] Connolly, 'Newton and God's Sensorium', p. 190

[72] See Henry and McGuire, 'Voluntarism and Panentheism', pp. 2–6.

[73] Ibid., p. 3

the soul and the sensitive substance was 'by no means Newton's own idea'; rather, it was 'the standard view of the sensorium and how it interacted with the soul'.[74]

And this leads to their second error, for they claim that the *sensorium* is identical with common sense. To support their previous claim of a 'standard view' of the *sensorium's* interaction with the soul, they first quote a short passage from a 1698 book and then suggest that Descartes's *Treatise on Man* contained 'something very similar'.[75] To support this claim, they quote a passage from it to the effect that what should be taken as ideas are figures traced on the surface of the pineal gland, where 'the seat of imagination and "common" sense is located'[76] and where the rational soul, once it is united to the body machine, will consider directly.

According to Henry and McGuire, this passage means that 'the soul considers the forms or images in the sensorium ("the common sense")'.[77] Indeed, they repeated this claim by asserting that the 'reference to the "common" sense is Descartes's way of referring to the sensorium, also known as the *sensus communis*'.[78] But as previously indicated, Descartes did not use the term *sensorium*. For him (and for many others), the term 'common sense' denoted a *faculty* of the soul, a faculty that he called by the name 'animal spirits'. Moreover, there is no soul in the Cartesian body machine, because his mechanistic physiology of reflex treated automatic actions only.

In both papers commented on above, the three authors assumed that Descartes's ideas are the appropriate historical background for understanding Newton's ideas. Indeed, this assumption is common in much of the secondary literature, as may be instanced in a paper refreshingly free from undocumented speculation. Published in 2014, and devoted to Newton's physiology of vision, its author, Philippe Hamou, claimed that Newton 'aimed at constructing an interpretation of the Cartesian paradigm that on many points takes issue with Descartes' way of fleshing out the paradigm'.[79] To provide evidence for this claim, he singled out the early essay in Newton's notebook, 'Of Colours', mentioned previously; and he described this essay as offering a 'Cartesian-like' cerebralist account of perception, even though it did not completely adhere to the Cartesian 'legacy'. But if Hamou's statement were true—that Newton located 'the soul's dwelling place' in the human *sensorium*[80]—then this would point to a major difference between his concept of perception and that of Descartes. However, as pointed out above, in the essay Newton does not use

[74] Ibid., pp. 3–4

[75] Ibid., p. 4

[76] The quote is from a modern compilation in which it is unlikely that the compilers placed double quotation marks around the word 'common'. That it must be the authors' addition seems probable, since it is not found in Descartes's *Treatise* either in its original Latin or French versions or in the full translation of, and commentary on this work by T. S. Hall (see *infra* References). It is probable, therefore, that the double quotation marks were added by the authors themselves.

[77] Henry and McGuire, 'Voluntarism and Panentheism', p. 4

[78] Ibid., p. 19 n.13

[79] Hamou, Vision, Color, and Method, p. 32

[80] Ibid., p. 89

the term 'soul'. Moreover, in Newton's writings intended for a public audience, he does not provide a specific location for the soul; rather, he only indicates that this entity is located in some undisclosed place in the anatomical space of the brain.

Unlike the forementioned three papers, Liam Dempsey, in a paper published in 2006, cast aside Descartes as an appropriate background for understanding Newton's ideas. Instead, he claimed that Newton accepted 'a sort of mind-body monism, one which defies neat categorization, but which clearly departs from Cartesian substance dualism'.[81] In arguing his case, he did not involve the human *sensorium*, even though he mentioned it, as well as the divine *sensorium* in his paper. Instead, he claimed that the medium between mind and body is an electric spirit, a medium first mooted by Newton in the 'General Scholium' of the second, 1713 edition of *Principia mathematica*. Then, towards the end of his paper, Dempsey made a further claim, namely, that for Newton, the mind and the electrical spirit are 'quite similar in nature'.[82]

Unfortunately, not having read the pertinent literature, Dempsey was unaware that the electric spirit is one of Newton's hypothetical corporeal substances. Consequently, if the electrical spirit is similar to the mind, then the latter is corporeal; and if the latter is corporeal, then Newton is a materialist, that is to say, an atheist, like his (misjudged) bugbears, Descartes and Hobbes.[83] But in attempting to provide an appropriate theological background for his claim, Dempsey ignored Newton's well-known commitment to theism. It is therefore worth recalling the words of the seventeenth-century Cambridge Platonist, Ralph Cudworth, who, after producing extensive arguments against a variety of atheisms, pointed out that:

> ...this *Controverssie* betwixt Theists and Atheists, may be yet more Particularly Stated, from the *Idea* of God, as including *Mind* or *Understanding* in it *Essentially, Viz.* Whether *Mind* be *Eternal and Unmade*, as being the Maker of all; or else Whether *all Mind* were it self *Made or Generated*, and that out of Sensless Matter? For according to the Doctrine of the Pagan Theists, Mind ... [was the] *Oldest of all things, Senior to the World and Elements; and by Nature hath a Princely and Lordly Dominion over all.* But according to those *Atheists*, who make *Matter* or *Body* devoid of all *Life* and *Understanding*, to be the First *Principle*, Mind must be ... *A Post-Nate thing,* Younger than the world; a Weak, Umbratil, and Evanid Image, and next to Nothing.[84]

---

[81] See Dempsey, 'Written in the Flesh', p. 420.

[82] Ibid., p. 438

[83] For Descartes, see Newton, 'De gravitatione', pp. 142–3, where he asks rhetorically: 'If we say with Descartes that extension is body, do we not manifestly offer a path to Atheism'. For Hobbes, see Westfall, *Never at Rest*, p. 534, who quotes from Newton's letter of apology to Locke 'for representing that you struck at ye root of morality in a principle you laid down in your book of Ideas & designed to pursue in another book & that I took you for a Hobbist'. For the principle in question, see Locke, *An Essay Concerning Human Understanding*, pp. 538–43, where he raised certain problems that instance the limits of human knowledge, e.g. whether matter can think or whether the soul is a thinking incorporeal substance.

[84] Cudworth, *The True Intellectual System*, p. 729. For a brief treatment of the theory of spiritual knowledge associated with the tradition of Platonic 'anthropology', see also Dockrill, 'The Heritage of Patristic Platonism'.

In the inquiry that follows, I recognise the impact of Cartesianism when evidence allows. But I eschew methods that regard history through the lens of post-Cartesian interests, terminology and concepts. Instead, I try to understand and treat the interests, terminology and concepts of Newton, Willis and Briggs as authentic for their time, even if some of their terms and concepts are no longer valid. While Newton did not always have logical reasons for disliking Descartes and others,[85] I find it hard to believe that such a rigorous thinker would have imbibed, without critical sifting, an eclectic mix of philosophical traditions or influences, especially because, as is well known, he sought to establish methods that he hoped would deliver certainty in natural philosophy. Nevertheless, I acknowledge that such an epistemological project does carry some ontological commitments in the form of general assumptions about the structure of an object of knowledge, which in the case of this inquiry is Newton's *sensorium* concept, the underlying model that provides the meaning for both the human and divine parts of his concept.

Regarding Newton's religious beliefs and interests, which were wide-ranging, these are important only insofar as they have relevance to the problem of the divine *sensorium*. Of course, this constraint rules out topics such as his heterodoxy, his interest in church history and prophecies and his manuscripts that provide data for such topics, as well as any secondary discussion of these topics by increasing numbers of commentators. As for the question posed by Peter Harrison concerning whether Newton's theology is voluntarist, I attempt to answer it by providing an interpretation of Newton's agent-causation concept. If this interpretation is correct, then it will be clear that in the case of humans, as well as God, Newton does not subordinate intellect to will and, hence, cannot be described as a voluntarist.

Finally, to determine in Part IV what Newton means by the divine *sensorium*, it was necessary to enter the murky waters of a problem that has been the long-time focus of many commentators, namely, the problem of space, because Newton several times correlates space with the divine *sensorium*. However, since my interpretation of the latter is based on evidence cumulatively adduced in the first three parts of this inquiry, as well as on Newton's method of reasoning from experience, I hope to have accounted for the problem of space, as well as for the relationship between the two parts of his *sensorium* concept—human and divine—and, hence, for the whole of his theoretical construct.

***

[85] For example, the church father, Athanasius; see Wiles, *Archetypal Heresy*, pp. 87, 90–2; see also Westfall, *Never at Rest*, pp. 314–5, 344–5, *et passim*. Newton, echoing the myth of corruption from a pristine state, could claim that the mature Athanasius corrupted church doctrine with 'metaphysical opinions', which, in turn, led to corruptions of Scripture relating to the doctrine of the Trinity, a doctrine of the tradition of post-Nicene Western Christianity, the inheritor of which was Roman Catholicism. Newton, however, was motivated, first, by his desire to return to the tradition of ante-Nicene Christianity as practised during the second and third centuries (when the church was Arian) and, second, by his detestation of what he believed were idolatrous practices of the Roman church (e.g. the worship of saints). Indeed, for him, idolatry was *the* fundamental sin that corrupted not only religion but also natural philosophy; see Buchwald and Feingold, *Newton and the Origin of Civilization*, pp. 153–5.

For reasons beyond my control, this inquiry concludes nearly 50 years as a scholar, a life that has been increasingly difficult to maintain following an accident in 2013 and its subsequent life-changing consequences. *Quicquid corrigere est nefas*. During the period of work on this book, Mordechai Feingold offered continuing encouragement and advice by checking up on my progress, reading drafts and revisions and alerting me to new research relating to my inquiry, whereas Alan E. Shapiro shared his vast knowledge of Newton's optical writings by reading some of my drafts, providing helpful comments on them and offering suggestions for further reading. I am deeply grateful to both scholars for their generosity in taking time away from their own projects and commitments. Acknowledgements are due also to John Gascoigne, for reading an early version of the Introduction; the now late David Fairservice, for his translations of the Latin texts; the National Library of Australia in Canberra, for their assistance in obtaining copies of articles otherwise unavailable to me; and, my husband, Michael Kassler, for his loving support that has made possible the completion of this inquiry into Newton's *sensorium* concept.

Contents

List of Figures

Part I

I
The *Sensorium* in Newton's Texts

1.1 Preliminary Remarks

The following twelve texts either mention, or provide a context for Newton's use of the term '*sensorium*' in relation to humans as well as to God. Boldface is used as a means of drawing attention not only to the term, but also to other words and phrases that are important for the interpretations made during this inquiry. The texts themselves are set out in chronological order. Each text is numbered to facilitate cross reference, after which its bibliographical source is identified. When required, comment takes the form either of footnotes or remarks at the conclusion of a text. Both kinds of comment, though in different ways, concern textual matters, including meaning of terms, some of which are now rare or obsolete but which were current in Newton's day. The chief source for the latter kind of comment is the multi-volumed *Oxford English Dictionary*, hereafter cited as *OED*.

When possible, both kinds of comment also include identification of Newton's own sources either from hints in his texts or from two publications that identify books he owned or could have read. One source is the catalogue of Newton's own library[1]; the other, the catalogue of the library of his academic colleague at Cambridge University, Isaac Barrow.[2] According to Mordechai Feingold, Barrow was liberal not only in lending and even gifting his books to friends, but also in granting access to his library, an access that Newton had 'ever since his undergraduate days'.[3]

As indicated in the Introduction, the texts are extracted from writings that Newton intended for a public audience. In chronological order these are as follows: a 1673 a paper published in *Philosophical Transactions of the Royal Society*; a set of papers

[1] See Harrison, *The Library of Isaac Newton*.

[2] See Feingold, 'Isaac Barrow's Library'.

[3] Ibid., p. 336.

J. C. Kassler, *Newton's* Sensorium*: Anatomy of a Concept*, Archimedes 53,
https://doi.org/10.1007/978-3-319-72053-1_1

sent by Newton to the Royal Society and read to the Fellows at consecutive meetings between 1675 and 1676; the first 1704 edition of the *Opticks*, hereafter cited as *Opticks (1)*; its expanded 1706 Latin translation, hereafter cited as *Optice*; and, for the last three texts (X–XII), the original version of the fourth 1730 edition of the *Opticks*, hereafter cited as *Opticks (2)*.

Note, however, that between the first 1704 and fourth 1730 English editions, there were two other editions. One, published in 1717 and re-issued in 1718,[4] was identified as 'The Second Edition, with Additions'; the other, published in 1721, was identified as 'The Third Edition, Corrected'. To discover whether these two editions contain emendations (additions and/or deletions) or corrections to the texts presented here, I have consulted the study of Alexandre Koyré for changes to the Queries that conclude *Opticks (1)*.[5] For changes to other parts of *Opticks (1)*, I have relied on the modern version of the fourth 1730 edition of *Opticks (2)* published by Dover in 1952, hereafter cited as *Opticks (C-R)*.

According to Alan Shapiro, Newton began composing *Opticks (1)* in Latin, reaching as far as 'the better part of Book I before he decided to rewrite it in English'.[6] In the subsequent English editions, therefore, much of Book I is his own free English translation of his Latin draft.[7] With the exception of the concluding Queries, it seems that no systematic comparison has been made between the various transformations of *Opticks (1)*. However, Shapiro did notice that in the 1706 *Optice*, the translator, Samuel Clarke, substituted the Latin word '*imago*' for the English word 'spectrum', even though Newton had introduced the latter word as a technical term in optics.[8]

Finally, I am not presenting a thesis to be evidenced and demonstrated; rather I am pursuing a process of discovery that follows from the data. For this reason the data, including both kinds of comment mentioned above, is provided before the inquiry proper begins in Part II. In this way I hoped to involve readers in the discovery process, even if they bring different areas of expertise or different assumptions to their reading. Moreover, to avoid omitting information important not only for readers but also for this inquiry, I sought to provide appropriate amounts of context from the passages in Newton's writings that include the term '*sensorium*'. For his complete *sensorium* concept cannot be elucidated either by the practice of glossing the term '*sensorium*[9] or by the practice of extracting from his writings a sentence or less as a basis of interpretation.[10]

---

[4] See Wallis and Wallis, *Newton and Newtoniana*, p. 130.

[5] See Koyré, 'Les Queries de l'Optique'.

[6] See Shapiro, 'The Spectre of Newton's "Spectrum"', according to whom Newton titled the Latin draft '*Fundamentum opticæ*' (personal communication).

[7] Ibid., p. 190.

[8] Ibid., pp. 189–90. The Latin translation was seen through the press by Abraham de Moivre; see Westfall, *Never at Rest*, p. 648.

[9] See *supra* Introduction, p. v and nn.1 and 2.

[10] See *supra* ibid., pp. xxvii, xviii.

1.2 The Data

Text I

'An Extract of Mr. Isaac Newton's Letter, written to the Publisher from Cambridge April 3 1673 concerning the Number of Colors, and the Necessity of mixing them all for the production of White....', *Philosophical Transactions of the Royal Society* (6 October 1673), VIII, No. 97 (pp. 6108–11) p. 6011[11]

> ...[in my first letter[12]] I wonder'd, how Telescopes could be brought to so great perfection by Refractions which were so Irregular. But, to take away the difficulty, I must acquaint you ... that ... although Light, which passes through the middle parts of the Glass, were wholly intercepted, yet would the remaining light convene infinitely more dense at the due points, than at other places. And by this excess of Density, the Light, which falls *in* or invisibly *near* the just point, may, I conceive, **strike the *sensorium*** *so vigorously*, that the impress of the weak light, which errs round about it, shall, in comparison, not be strong enough to be animadverted, or to cause any more sensible confusion in the Picture than is found by Experience.

In the previous part of his letter, Newton's concern is his theory of colour, whereas the final part treats the difficulty mentioned in the passage quoted above. To the best of my knowledge, this is the earliest reference to a *sensorium* in the writings that Newton made public.

Text II

'An Hypothesis explaining the Properties of Light, discoursed of in my several Papers',[13] read to the Fellows of the Royal Society between December 1675 and February 1676[14]

> [pp. 182–5]...that puzzling problem; ***By what means the muscles are contracted and dilated to cause animal motion***, *may receive greater light from hence* [the æther] *than from any means men have hitherto been thinking on.* For, if there be **any power**[15] **in man to**

[11] The 'Extract' is part of a reply to Christiaan Huygens; for a facsimile of the entire letter, see Cohen and Schofield (eds.), *Isaac Newton's Papers*, pp. 143–6.

[12] See 'A Letter of Mr. Isaac Newton ... containing his New Theory about Light and Colours....', *Philosophical Transactions of the Royal Society* (19 February 1671/2), VI, No. 80, pp. 3075–87, which is reproduced in facsimile in Cohen and Schofield (eds.), *Isaac Newton's Papers*, pp. 47–59.

[13] I.e., his papers published previously in the *Philosophical Transactions of the Royal Society*, for which see the facsimiles in Part I of Cohen and Schofield (eds.), *Isaac Newton's Papers & Letters*.

[14] The manuscript was afterwards published in volume 3 of Thomas Birch, *The History of the Royal Society* (London, 1757). I have used the facsimile reprint in Cohen and Schofield (eds.), *Isaac Newton's Papers*, pp. 177–235 (which includes Birch's pagination). According to Shapiro, *Fits, Passions, and Paroxysms*, p. 72 n.59, the version of the paper in the Journal Book of the Royal Society (i.e., Birch's version) was published in Newton, *Correspondence*, vol. 1, pp. 362–86, with supplements and corrections from Newton's autograph copy in Cambridge University Library Add. 3970: ff.538–47. Note also that on 25 January 1675/6, Newton wrote to Robert Boyle regarding some corrections to Text II; see Cohen and Schofield (eds.), *Isaac Newton's Papers*, p. 74.

[15] OED *sb.*[1] 3, Of inanimate things: Active principle or property; capacity of producing an effect (1592–1871).

condense and dilate at will[16] **the æther, that pervades the muscle, that condensation or dilation must vary the compression of the muscle, made by the ambient æther, and cause it to swell or shrink** accordingly. For though common water will scarce shrink by compression, and swell by relaxation, yet (so far as my observation reaches) spirit of wine and oil will; and ... **animal juices do the same**. And as for their various pression[17] by the ambient æther, it is plain that that must be more or less accordingly as there is more or less æther within, to sustain and counterpoise the pressure of that without.... **To vary the compression of the muscle therefore, and so to swell and shrink it**, there needs nothing but to change the consistence of the included æther; and a very little change may suffice, if the spring of æther be supposed very strong, as I take it to be many degrees stronger than that of air.

Now for the changing the consistence of the æther[18]; ... I shall not stick to tell you how I think it may be done.

First then, **I suppose, there is such a spirit;** that is, **that animal spirits are** neither like the liquor, vapour, or gas[19] of spirit of wine; but **of an ætherial nature, subtil enough to pervade the animal juices**, as freely as the electric, or perhaps magnetic, effluvia do glass. **And to know, how the coats**[20] **of the brain, nerves, and muscles, may become a convenient vessel to hold so subtil a spirit, you may consider,** how liquors and spirits are disposed[21] to pervade or not pervade things on other accounts than their subtilty. ... So some fluids, as oil and water, though their parts are in freedom enough to mix with one another, yet by some secret principle of unsociableness they keep asunder; and some, that are sociable, may become unsociable, by adding a third thing to one of them, as water to spirit of wine, by dissolving salt of tartar in it.[22] The like unsociableness may be **in** ætherial natures.... And on this ground, if **the ætherial vital spirit in a man** be very sociable to **the marrow**[23] **and juices,** and unsociable to **the coats of the brain, nerves, and muscles, or**

[16] OED *sb*[1] III. Special uses, 15: According to one's volition or choice (13..–1888).

[17] OED 1, rare: The action of pressing; pressure (1662–1756). The 1662 citation is to the second edition of Boyle, *New Experiments Physico-Mechanicall, Touching the Spring of the Air*, in which the author attempted not only to give a qualitative demonstration of the air's elasticity ('spring' or resistance to compression) but also to reduce this phenomenon to a qualitative law by correlating two variables initially termed 'spring' and 'density' and subsequently altered to 'pressure' and 'weight'. For the concept of the elasticity of air in the seventeenth century, see Webster, 'The Discovery of Boyle's Law'. According to Westfall, *Never at Rest*, p. 376, Boyle's *Spring of Air* was one of the first books on natural philosophy read by Newton. If so, he would have read the first 1660 edition.

[18] Here, Newton identifies three different suppositions concerning how the soul has a power to change 'at will' the consistence of the æther (and, hence, the swelling and shrinking of the muscle). See further *infra* Pt.II.2.1, p. 41.

[19] The term 'gas' was coined by the Dutch physician, J. B. van Helmont to mean an occult principle contained in all bodies and conceived by him as an ultra rarefied condition of water. For its English use, see OED *sb.*, obs. (1658–1743). For its meaning as any aeriform or completely elastic fluid, the first citation in OED is 1779.

[20] I.e., what later in this text he refers to as stiff 'ætherial superficies' or surfaces.

[21] OED *ppl.* 2b, obs.: Have a (particular) bodily constitution; constituted (c.1386–1694); thus, 5: Have a physical tendency (c.1380–1886).

[22] The secret principle referred to in this sentence is a natural agent, the chemical action of which mediates between two passive substances, whereby they become sociable or unsociable one to another. For the action that Newton, in the following paragraph, refers to as 'mediation', see *infra* Pt.II.2.1, p. 42. See also the introduction to Newton's chemical papers in Cohen and Schofield (eds.), *Isaac Newton's Papers*, pp. 241–8.

[23] I.e., that which is innermost; the marrow or medullary substance under the ætherial surfaces called coats.

to any thing lodged in the pores of those coats, [then] **it** [the ætherial vital spirit] **may be contained thereby, not withstanding its subtilty....**

In the next place, **for knowing how this** [ætherial vital] **spirit may be used for animal motion, you may consider,** how some things unsociable are made sociable by the mediation of a third.[24] ... And in like manner [as a solvent], **the ætherial animal spirit in a man may be a mediator between the common æther and the muscular juices, to make them mix more freely**; and so, **by sending a little of this** [ætherial animal] **spirit into any muscle, though so little as to cause no sensible tension of the muscle by its own force**; yet, by rendering **the juices** more sociable to the common external æther, it may cause that æther **to pervade the muscle** of its own accord in a moment more freely and copiously than it would otherwise do, and **to recede** again as freely, so soon as **this mediator**[25] **of** sociableness is retracted. **Whence**, according to what I said above, **will proceed the swelling or shrinking of the muscle, and consequently the animal motion depending thereon**.

Thus may therefore **the soul, by determining this ætherial animal spirit or wind into this or that nerve**, perhaps with as much ease as air is moved in open spaces, **cause all the motions we see in animals**: for the making which motions strong, it is not necessary, that we should suppose **the æther within the muscle** very much condensed or rarified by this means, but only that its spring is so very great, that a little alteration of its density shall cause a great alteration in the pressure. **And what is said of muscular motion, may be applied to the motion of the heart, only with this difference, that the** [ætherial animal] **spirit is not sent thither, as into other muscles, but continually generated there by the fermentation of the juices, with which its flesh is replenished, and as it is generated, let out by starts**[26] **into the brain through some convenient ductus**[27] **to perform those motions in other muscles by impression, which it did in the heart by its generation**. For I see not, why **the ferment in the heart may not raise as subtil a spirit out of its juices, to cause these motions**, as rubbing does out of a glass, to cause electric attraction....

[pp. 185–6]

Hitherto I have been contemplating the nature of æther and ætherial substances by their effects and uses; and now I come to join therewith the consideration of light ... [which] I suppose is neither æther, nor its vibrating motion, but something of a different kind propagated from lucid bodies. ...[28] ...I suppose [light] consists of rays differing from one another in contingent circumstances, as bigness,[29] form,[30] or vigour[31]; like as the sands on a shore....[32] And further, I would suppose it [light] diverse, from the vibrations of the æther ...

---

[24] I.e., a solvent, as is indicated in the examples that follow, but not reproduced here. For his final version relating to dissolution and other chemical processes, see Newton, *Opticks (C-R)*, Qu. 31, pp. 375–406, pp. 376–89.

[25] I.e., the ætherial animal spirit.

[26] I.e., intermittently, as in the phrase, fits and starts; see OED *sb.*[2] 4c (transferred from 3: paroxysm): By irregular pulses or periods of action, at varying intervals, fitfully, spasmodically (1583–1884).

[27] I.e., a Latin form, in use during the seventeenth century, for duct; see OED 6a *Phys.*: A tube or canal in the animal body for the conveyance of bodily fluids (1667–1872).

[28] Here, Newton identifies three different suppositions concerning the nature of light, perhaps in parallel with his identification of three different suppositions concerning muscle action (see p. 6 n.18 below).

[29] I.e., magnitude. Newton also uses the same term with a different meaning; see p. 8 n. 37 below.

[30] I.e., due shape or figure

[31] OED *sb.* 3: Active force or strength as an attribute of natural agencies, conditions or qualities; intensity of operation (1445–1880); for the term 'vigour', see also continuation of this text [p. 192] below.

[32] In what follows, therefore, for the word 'light' read 'rays of light'.

[and] that light and æther mutually act upon one another, æther in refracting light, and light in warming æther; and that the densest æther acts most strongly.

[pp. 188–9]...I suppose, that the rays, when they impinge on the rigid resisting ætherial superficies, as they are acted upon by it, so they react upon it and cause vibrations **in** it ... and that these vibrations are propagated every way into both the rarer and denser mediums.... For it is plain by **the heat**, which light produces **in** bodies, that **it** is able to put their parts in motion, and much more **to heat** and put in motion the more tender æther; and it is more probable, that **it** [heat] communicates motion to the gross parts of bodies by the mediation of æther than immediately. ... Ætherial vibrations are therefore the best means by which such a subtile agent[33] as light can shake the gross particles of solid[34] bodies **to heat** them: and so supposing that light, impinging on a refracting or reflecting ætherial superficies, puts it into a vibrating motion, that physical superficies being by the perpetual appulse[35] of rays always kept in a vibrating motion, and the æther **therein** continually expanded and compressed by turns; if a ray of light impinge upon it [the superficies], while it is much compressed, I suppose it is then too dense and stiff to let the ray pass through, and so reflects it; but the rays, that impinge on it at other times, when it is either expanded by the interval of two vibrations, or not too much compressed and condensed, go through and are refracted. ...

Lastly, because without doubt **there are, in every transparent body, pores of various sizes**, ... æther stands at the greatest rarity **in the smallest pores**.... But where **the pores** exceed a certain bigness, I suppose the æther suits its density to the bigness of **the pore**, or to the medium **within it**....

[p. 192] ...I suppose, that as **bodies of various sizes, densities, or sensations**,[36] do by percussion or other action excite sounds of various tones, and consequently of various bigness[37]; so when the rays of light, by impinging on the stiff refracting superficies, excite vibrations **in** the æther, those rays, whatever they be, as they happen to differ in magnitude, strength or vigour, excite vibrations of various bigness; the biggest, strongest, or most potent rays, the largest vibrations; and others shorter, according to their bigness, strength, or power: **and therefore the ends of the capillamenta**[38] **of the optic nerve**[s], **which pave or face** [the back of] **the retina** [in each eye], **being such refracting superficies, when the rays impinge upon them, they must there excite these vibrations, which** vibrations (like those of sound in a trunk or trumpet[39]) **will run along the aqueous pores of crystalline pith**[40] **of the capillamenta through the optic nerves into the sensorium** (which light

[33] OED *sb* B.5 Of things: The material cause or instrumentality whereby effects are produced, but implying a rational contriver (1579–1878).

[34] OED *a.* I.3, Of material substances: Of a dense consistency (c.1532–1887) = Dense OED 1: Having its constituent particles closely compacted together (1599–1878).

[35] I.e., striking on the previously-named ætherial surfaces; see p. 6 n.20 above.

[36] I.e., 'bodies' possessing the property of sensitivity.

[37] I.e., vibration lengths, intervals or periods, a meaning not included in OED under the term 'bigness'.

[38] I.e., hair-like fibres or filaments, the ultimate ramifications in living matter; see further *infra* Pt.II.2.1, pp. 35, 37–8.

[39] For the same analogy, see Newton, 'Of Colours', p. 488.

[40] I.e., marrow.

itself cannot do) **and there, I suppose, affect the sense** [of seeing] **with various colours,** according to their bigness and mixture; the biggest with the strongest colours, reds and yellows; the least with the weakest, blues and violets; the middle with green, and a confusion[41] of all with white, much after the manner, that in the sense of hearing, nature makes use of ærial vibrations of several bignesses to generate sounds of divers tones; **for the analogy of nature is to be observed.**[42] And further, as the harmony and discord of sounds proceed from the proportions of the ærial vibrations, so may the harmony of some colours, as of golden and blue, and the discord of others, as of red and blue, proceed from the proportions of the ætherial [vibrations].

Newton's sequence of papers commences with a short introduction (pp. 178–9) in which he states, first, that 'all allowable hypotheses in their genuine constitution should be conformable to my theories' and, second, that if he were to assume an hypothesis, then 'it should be this, if propounded more generally, so as not to determine what light is, farther than that it is something or other capable of exciting vibrations in the æther'. After this introduction, he begins the hypothesis proper concerning the æther, which, he states, 'is compounded partly of a phlegmatic body of æther, partly of other various ætherial spirits, much after the manner, that air is compounded of the phlegmatic body of air intermixed with various vapours and exhalations' (pp. 179–84). He then applies this hypothesis to two main subjects: one, concerning the nature of light and its relation to the æther (pp. 184–90); the other, concerning his theory of colour (pp. 162–235).

In his superb treatment of 'An Hypothesis', Alan Shapiro traced the development of its two main parts from the 1660s to the publication of *Opticks (1).* He observed that this piece of writing is 'entirely consistent with Newton's methodology, which allowed hypotheses to be used to explain properties, provided they were based on empirical evidence and not mingled with more certain principles'.[43] He made the further observation that the æther 'played two, divergent roles' in Newton's natural philosophy 'that merged into one another'.

On the one hand, in a Cartesian manner he invoked the mechanical properties of an invisible, subtle fluid that acted only by contact to explain such physical phenomena as cohesion, electrical attraction, and the refraction and reflection of light. On the other hand, Newton also imagined the aether to be a source of activity in nature for such vital processes as fermentation, putrefaction, and generation, growth, and decay of living matter that could not, he felt, be explained by mere mechanical causes.[44]

Regarding the second role of the æther as a source of activity in nature, it should be noted that all the vital processes mentioned above by Shapiro were treated by Thomas Willis in his first publication, *Diatribæ duæ medico-philosophicæ, quorum*

---

[41] I.e., a mixture in which the distinction of the elements is lost by intermingling.

[42] For the meaning of this maxim, see *infra* Pt.IV.4.2, pp. 109–10.

[43] See Shapiro, *Fits, Passions, and Paroxysms*, pp. 72–85, p. 72.

[44] Ibid., pp. 73–4.

prior agit de fermentatione (London, 1659).[45] In addition, the tract includes three other features that deserve notice here. First, Willis set out his conception of five chemical 'principles', the most important of which, 'spirit', is described as highly subtle 'Ætherial Particles of a more Divine Breathing, which our Parent Nature hath hid in this Sublunary World, as it were, *the Instruments*[46] *of Life and Soul, Motion and Sense*'.[47] Second, in announcing that 'all things' are 'full of this Ætherial Substance', as well as 'full of Fermentation',[48] the tract anticipates what later became the foundation of Willis's physiology.[49] Third, he used terms such as consociate and dissociate to describe the association and separation of the respective chemical principles.[50] Accordingly, if Newton read this tract (and I believe he did[51]), the latter terms, as well as the contexts in which they occur, might have suggested to him the nature of the secret principle that mediates sociability and unsociability.[52] But this conjecture remains for others to explore.

As for the æther's first role, Shapiro pointed out that Newton adopted the underlying model of vibrations in the æther as early as 1666 and continued to utilise it 'decades after he had abandoned the aether itself'.[53] Text II, therefore, provides an early version of this mechanical model insofar as the visual sensory system is concerned. But for the purposes of our inquiry, the text is more important for the

---

[45] Two other tracts (on fevers and urines) were included in this publication, which in its Latin form was reissued in London in 1660, 1662 (2d edn, with additions) and 1677. There also were a number of continental issues, for which see Davis, *Circulation Physiology and Medical Chemistry*, p. 154 n.49.

[46] I.e., material causes.

[47] See Willis, *A Medical-Philosophical Discourse* [1659 tr. 1681], p. 3 (italics mine).

[48] From the Latin *ferveo*, meaning to boil.

[49] Willis, *A Medical-Philosophical Discourse* [1659 tr. 1681], pp. 12, 16; see also *infra* Pt.III.3.2, pp. 44–5.

[50] Willis, *A Medical-Philosophical Discourse* [1659 tr. 1681], pp. 1, 5–8, 16, 27, 38, 40–2, 49–51. Note, however, that terms used to express the boiling motion of fermentation (e.g., 'association' or 'congregation' and 'separation') were not uncommon; see, e.g., the 1644 theory of Thomas Hobbes in Kassler, *Inner Music*, p. 65.

[51] Barrow owned the first 1659 edition of Willis's book; see Feingold, 'Isaac Barrow's Library', (No. 1049), p. 360.

[52] See p. 6 n. 22 above. See also Newton, 'De motu sensatione Animalium' and 'De vita & morte vegetabili', which represents a draft appendix for the second 1713 edition of *Principia mathematica*. In the second text he treats the vital processes of generation, growth and decay of living matter (vegetable and animal) and uses the term 'attraction' rather than 'sociability'. He also uses the term, *flamma vitalis*, p. 81, which, due to the proto-chemical context of the manuscript, may refer to Willis's conception of a vital flame in the blood. If the approximate dates assigned to these two manuscripts are correct, then Newton seems unaware that in 1674 the physiologist, John Mayow, rejected Willis's conception, and it soon became *passé*; see Frank, Jr., *Harvey and the Oxford Physiologists*, p. 270.

[53] Shapiro, *Fits, Passions, and Paroxysms*, p. 72, referring to Newton, 'Of Colours'.

insight it provides into Newton's suppositions concerning animal motion. Indeed, this subject occurs in only one other, highly abbreviated (we might even say, Delphic) text, which re-introduces an ætherial medium (see Text X below).

Nevertheless, books in Newton's library point to an on-going interest in animal motion and, especially, the motive power that contracts the muscles, as will be obvious from the titles that contain such words as animal motion or muscles (including words beginning with the prefix 'myo').[54] These books were published after 1675/6, the date of Text II. But with access to the library of Isaac Barrow, Newton could have read a book, published anonymously in 1664, that proposed to explain the motive power that contracts the muscles.[55] And as a Fellow of the Royal Society, he would have had access to reviews of books in its *Philosophical Transactions* that treated muscle action in particular or as part of a general anatomy.[56]

However, one book in Newton's library, which does not have the words 'animal motion' or 'muscles' in its title, contains in the very first chapter an hypothesis concerning muscle action. The book in question is a 1668 continental edition of Willis's *Patholgiæ cerebri, et nervosi*, published in 1667,[57] the central thesis of which is that the motive power in the muscles is produced in the their fibres by a chemical reaction that involves mixing ('coupling') of two kinds of substances. One is a natural, phlegmatic ferment sent with the blood from the heart through the arteries; the other is an ætherial substance called 'animal spirits' sent from the brain through the nerves.[58] It is worth recalling, therefore, that according to Newton's hypothesis quoted at the outset of this comment, his æther is also compounded of two unobservable and, hence, hypothetical substances—one, 'a phlegmatic body of æther' and the other, 'various ætherial spirits'.

---

[54] See Harrison, *The Library of Isaac Newton*, (Nos. 172, 203, 248, 451 and 1676).

[55] The author, a physician, was William Croone, *De ratione motus musculorum* (London, 1664); see Feingold, 'Isaac Barrow's Library', (No. 287), p. 348.

[56] E.g., the review of the first book written by Nicolas Steno, *De musculis et glandulis observationem specimen* (1664), who suggested that the only way to solve the problem of muscle action was along mathematical and geometrical lines. He developed this suggestion in a paper published in 1667; see Bastholm, 'Niels Stensen's Myology'. Both the book and the paper were reviewed in *Philosophical Transactions of the Royal Society* (1667–68), 1–2, Nos. 10 and 32, pp. 176–8 and pp. 627–8.

[57] See Harrison, *The Library of Isaac Newton*, p. 262 (No. 1741). Note that Steno and Willis offered different explanations of muscle action; and, consequently, their concepts were frequently discussed by subsequent authors, who adopted, rejected or offered modifications to one or the other explanation.

[58] See Willis, *The Anatomy of the Brain* [1664 tr. 1681], p. 129, where he described the phlegmatic substance as a thick, 'dull', 'watery juice' and the ætherial substance as a 'spiritous liquor'; see also *infra* Pt.II.2.1, p. 43.

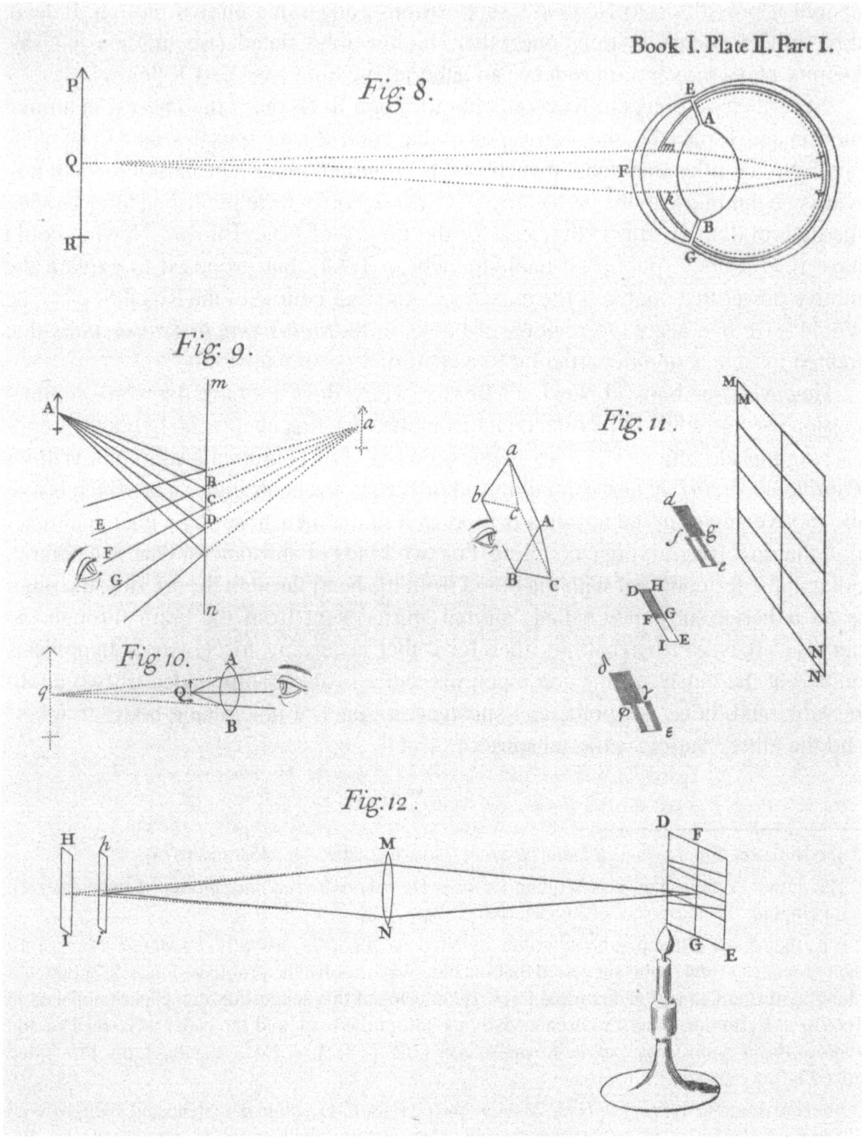

Fig. I.1 Optical process of image formation on the retina *from* Newton, *Opticks* 1704, Bk. I, Pt. I, Fig. 8 in Plate II

External object	PQR
Coat of the cornea	EFG
Pupil	*mk*
Crystalline humour	AB
Coat of the retina	shown in dotted lines

Text III
Opticks (1), Bk. I, Pt. I,
Axiom VII (pp. 9–11) p. 10 and Fig. 8 in Plate II

> ...when a Man views any Object PQR [see Fig. I.1], the Light which comes from the several Points of the Object is so refracted by **the transparent skins**[59] **and humours of the Eye**, (that is **by the outward coat** EFG called **the *Tunica Cornea***, and by **the crystalline humour** AB **which is beyond the Pupil** (*mk*) as to converge and meet again at so many Points **in the bottom of the Eye**, and **there** to paint the Picture of the Object **upon that skin** (called ***the Tunica Retina***) **with which the bottom of the Eye is covered**. For Anatomists when they have taken off from **the bottom of the Eye** that outward and most thick Coat called the *Dura Mater*,[60] can then see through the thinner Coats the Pictures of Objects lively painted thereon.[61] And these Pictures propagated by Motion **along the Fibres of the Optick Nerves into the Brain**, are the cause of Vision. For accordingly as these Pictures are perfect or imperfect, the Object is seen perfectly or imperfectly.

The remainder of the paragraph concerns imperfect vision from jaundice or old age and the effects of spectacles. Although the above extract contains no mention of a *sensorium*, yet it is of importance for this inquiry for reasons that will appear later on.[62]

Text IV
Opticks (1), Bk. I, Pt. II,
Theor. II (pp. 87–91) pp. 90–1

> **In them** [the rays of light] there is nothing else than a certain power[63] and disposition[64] to stir up **a sensation of this or that Colour**. For as sound in a Bell or musical String, or other sounding Body, is nothing but a trembling Motion, and in the Air nothing but that Motion propagated from the Object, and **in the Sensorium** 'tis **a sense of that Motion** under the form[65] of sound; so Colours in the Object are nothing but a disposition to reflect this or that sort of rays more copiously than the rest; **in** the rays they [the colours] are nothing but their dispositions to propagate this or that Motion **into the Sensorium**, and **in the Sensorium** they are **sensations of those Motions** under the forms of Colours.

[59] I.e., coats.

[60] I.e., the fibrous membrane forming the outermost covering of the brain.

[61] In 1625 Christoph Scheiner, recorded an experiment in which he removed the backs of human and animal eyes and observed the formation of the image on the retina. In 1637 Descartes made the experiment and described it fully but without naming Scheiner. See Boring, *Sensation and Perception*, p. 223; Crombie, *Science, Optics and Music*, pp. 293, 424; and Hall in Descartes, *Treatise of Man*, p. 51–2 n.90.

[62] See *infra* Pt.II.2.1, pp. 31–3.

[63] OED *sb.*[1] B. 3, Of inanimate things: Active principle or property; capacity of producing an effect (1592–1871).

[64] OED II. 10, obs.: Physical condition or state (c.1400–1732).

[65] OED *sb.* I. 4c: The real objective condition on which a sensible quality of a body depends for its existence, and the knowledge of which enables it to be freely produced (1605). For this meaning, which is a modification of scholastic usage, see Francis Bacon, *Of the Advancement of Learning*, Bk. II, vii.§5.

Text III is extracted from the 'Definition' that concludes Prop. II, Theor. II. This proposition states that 'All homogeneal Light[66] has its proper Colour answering to its degree of refrangibility, and that Colour cannot be changed by reflexions and refractions.' In the 'Definition' Newton points out that when he uses terms such as 'red-making', he is speaking vulgarly, not philosophically and properly, because, rays themselves are not coloured but have 'dispositions' for colour making.

In Text II Newton left rays of light undefined; but at the very outset of *Opticks (1)* he provided a short definition as follows:

> By rays of Light I understand its least Parts, and those as well as Successive in the same Lines as Contemporary[67] in several lines. ... The least Light or part of Light, which may be stopt alone without the rest of the Light, or propagated alone, or do suffer any thing alone, which the rest of the Light doth not or suffers not, I call a Ray of Light.[68]

The expression 'least parts' implies that rays are unobservable corporeal substances, a property of which is that they cannot be further divided or broken into pieces, that is to say, they are indivisible and, hence, absolutely hard atoms.[69] If this is the case, then, according to Text IV, least parts have two other properties, namely, 'powers' and 'dispositions', although these are left undefined. Indeed, regarding the latter property, Newton explicitly states that he does not 'here enquire' whether the disposition is 'a circulating or a vibrating motion of the ray, or of the medium, or something else'. And his reason is that he is content 'with the bare discovery, that the rays of Light are by some cause or other alternately disposed to be reflected or refracted for many vicissitudes',[70] that is to say, to change from one physical state to another as part of a constant series of changes ('vicissitudes').

Further suggestions concerning the atomic rays may be found in two of the new Queries added to the 1706 *Optice*, where Newton conceives 'least parts' as small bodies ('corpuscula') emitted from shining substances,[71] as inert and mobile, as perfectly hard ('perfecte dura') and impenetrable (therefore, not porous); and as original ('primigeniæ') because brought into existence by God.[72] He also points out that nothing is more requisite for producing all the variety of colours, and degrees of refrangibility, than that the rays of light be small bodies of different sizes. And

---

[66] I.e., light whose rays are all alike refrangible; see Newton, *Opticks (1)*, Defin. VII, p. 3.

[67] I.e., concurrent.

[68] See Newton, *Opticks (1)*, Bk. I, Defin., I, pp. 1–2.

[69] See Shapiro, 'Newton's Definition of a Light Ray', p. 209, who pointed out that, unlike Newton's previous definitions, that in *Opticks (1)* is definitely in terms of a light corpuscle or atom.

[70] See Newton, *Opticks (1)*, Bk. II, Pt.III, Prop. XII, p p. 78–81, pp. 80–1.

[71] See Newton, *Optice*, Qu. 21, p. 315.

[72] See ibid., Qu. 23, pp. 335–6, 343–4, 347. On the last page, Newton writes: 'Deum creare Materiæ particulas'. The literal meaning of 'creare' is 'to bring into existence' and by transference, its meaning is 'to give use to, occasion, cause'. Cf. Texts VIII, IX and IX below, where Newton does not use the term 'creare'.

further, he points out that nothing more is requisite for putting rays of light into 'fits' of easy reflection and easy transmission than that they be small bodies, which by their attractive powers, or some other force, stir up vibrations *in* what they act upon, which vibrations being swifter than the rays, overtake them successively, and agitate them so as by turns to increase and decrease their velocities and thereby put them into those fits.[73] The final clause in the foregoing sentence refers to Newton's theory of fits, a theory he set forth in *Opticks (1)* directly after declining to specify what he meant by dispositions of the rays.[74]

In a detailed account of this theory and its development, Alan Shapiro has suggested that Newton drew the term 'fits' from the semantic field of medicine, in which it was used as a synonym for intermittent fevers (agues) and treated as a disposition—physical condition or state.[75] As an instance, he mentioned an English translation, published in 1683, of a Latin tract on fevers published in 1659 by Thomas Willis. Although it is possible that Newton could have read one or the other of these versions, he did not own a copy of either of them. But he did own a book that seems a better source for Newton's term 'fits'. This is the previously mentioned 1668 continental edition of Willis's 1667 book on 'Convulsive *Pathology*', namely, *Pathologiæ cerebri, et nervosi* (see comment to Text II above).

In the first chapter of this book, Willis attempted to distinguish spasms from convulsions by considering them as different distempers, the one (spasm) as 'a continual Convulsive Cramp' and the other (convulsion) as 'A *Convulsive motion in general*'.[76] But in the rest of the book, this distinction is somewhat muddied, because the terms 'fits', 'spasms', 'paroxysms' and 'convulsions' seem to be used almost interchangeably to describe various convulsive diseases that include, in addition to intermittent fevers, a whole range of recurrent dispositions, of which the classic instance is the epileptic fit that Willis considered in detail in his second and third chapters.[77] Moreover, to treat the different symptoms of convulsive diseases, he, like the physicians of his day, determined a number of different kinds of causes (e.g., evident, conjunct, etc.), one of which, the 'procatartick'[78] was an external and occasional cause such as the previous disposition of a patient.[79]

---

[73] Ibid., Qu. 21, pp. 317–8.

[74] See Newton, *Opticks (1)*, Bk. II, Pt.III, Prop. XIII, pp. 81–6.

[75] Shapiro, *Fits, Passions, and Paroxysms*, pp. 180–1. Concerning the term 'fit', see also OED *sb.*[2] 3a: A paroxysm, or one of the recurrent attacks of a periodic or constitutional ailment (1547–1855).

[76] Willis, *An Essay of the Pathology of the Brain* [1667 tr. 1684], p. 1.

[77] Ibid., pp. 11–22.

[78] OED *a.* obs. (1603–1834).

[79] See Willis, *An Essay of the Pathology of the Brain* [1667 tr. 1684], p. 4: 'Convulsions, whose paroxisms [*sic*] are more grievous, and stay longer, or are oftener repeated, seem altogether to depend on a procatartick cause, or a previous disposition, and to arise from some other conjunct cause, besides irritation.'

Text V
Opticks (1), Bk. I, Pt. II,
Prop. V, Theor I (pp. 98–117)
Exper. X (pp. 99–105) pp. 102–4, and Exper. XII (pp. 106–8) p. 106

> Lastly, I made an Instrument ... in fashion of a Comb, whose Teeth being in number sixteen were about an Inch and an half broad, and the intervals of the Teeth about two Inches wide. Then by interposing successively the Teeth of this Instrument near the Lens, I intercepted part of the Colours by the interposed Tooth, whilst the rest of them went on through the interval of the Teeth to the Paper ... and there painted a round solar Image. ... [Afterwards] I caused ... all the Teeth to pass successively over the Lens, and when the motion was slow, there appeared a perpetual succession of the Colours upon the Paper: But if I so much accelerated the motion, that the Colours by reason of their quick succession could not be distinguished from one another, the appearance of the single Colours ceased ... but from a confusion[80] of them all there arose one uniform white Colour. Of the Light which now by the mixture of all the Colours appeared white, there was no part really white. One part was red, another yellow, a third green, a fourth blue, a fifth purple, and every part retains its proper Colour till it **strike the Sensorium. If the impressions follow one another slowly, so that they may be severally perceived, there is made a distinct sensation of all the Colours one after another in a continual succession.** But **if the impressions follow one another so quickly that they cannot be severally perceived, there ariseth out of them all one common sensation**, which is neither of this Colour alone nor of that alone, but hath it self indifferently to 'em all, and this is **a sensation of** whiteness. By the quickness of the successions **the impressions of** the several Colours **are confounded**[81] **in the Sensorium**, and out of that confusion ariseth **a mixt sensation**. If a burning Coal be nimbly moved round in a Circle with Gyrations continually repeated, the whole Circle will appear like fire; the reason of which is, that **the sensation of** the Coal ... in the several places of that Circle **remains imprest on the Sensorium**, until the Coal return again to the same place. And so in a quick consecution of the Colours **the impression of** every Colour **remains in the Sensorium**, until a revolution of all the Colours be compleated, and that first Colour return again. **The impressions therefore of** all the successive Colours **are at once in the Sensorium, and joyntly stir up a sensation of** them all; and so it is manifest by this Experiment, that **the commixt impressions of** all the Colours do **stir up and beget a sensation of** white, that is, that whiteness is compounded of all the Colours.
>
> **And if the Comb be now taken away, that all the Colours may at once pass from the Lens to the Paper, and be there intermixed,** and together reflected thence to the Spectators Eyes; **their impressions on the Sensorium being now more subtily and perfectly commixed there, ought much more to stir up a sensation of** whiteness.

> ...when the Comb was so quick [in its reciprocal motion up and down], that the Colours could not be distinguished from one another, the whole Paper **by their confusion and mixture in the Sensorium** appeared white.

Prop. V, Theor. IV (pp. 98–117) states: 'Whiteness and all grey Colours between white and black, may be compounded of Colours, and the whiteness of the Sun's

[80] I.e., mixture in which the distinction of the elements is lost by intermingling (as in Text II and p. 9 n.41 above).

[81] OED *v.* 6: To mix up or mingle so that the elements become difficult to distinguish.

Light is compounded of all the primary Colours mixt in a due proportion.'[82] To prove the latter assertion regarding the sun's light, Newton describes a number of his experiments, two of which utilise a comb, from which Text IV is extracted.[83]

These experiments, which were performed in a dark chamber, represent a development of one that Newton described in 1672,[84] when instead of a comb, he used a toothed wheel

> ...with many Spoaks or coggs in its perimeter, whose Interstices and thicknesses may be equal and of such a largeness, that, if the Wheel be interposed between the Prisme and the white concourse of the colours, one half of the Colours may be intercepted by a spoake or cogg, and the other half pass through an interstice. The Wheel being in this posture, you may first turn it slowly about, to see all the colours fall successively on the same place of the paper, held at their aforesaid concourse; and if you then accelerate its gyration, until the Consecution of those colours be so quick, that you cannot distinguish them severally, the resulting colour will be a Whiteness perfectly like that, which an un-refracted beam of Light exhibits, when in like manner successively interrupted by the spoaks or coggs of that circulating Wheel. And that this *Whiteness* is produced by a successive Intermixture of the Colours, without their being assimilated, or reduc'd to any Uniformity, is certainly beyond all doubt....

In his description of the toothed-wheel experiment, Newton did not address the problem of sensory perception. But in his description of the comb experiments, he does address this problem. For from what he has repeatedly observed during these experiments, he uses as generalisations for what cannot be observed, namely, that which happens *on* and *in* the *sensorium*. This method of reasoning from experience is an instance of what Newton, in Text II, referred to as the 'analogy of nature'.[85]

Text VI
Opticks (1), Bk. I, Pt. II,
Prop. VII, Theor. V (pp. 118–20) pp. 118–9

> It has been proved also, that when the several sorts of rays are mixed, and in crossing pass through the same space, they do not act on one another so as to change each others colorifick[86] qualities, (Exper. 10, *Lib.* 2) but **by mixing their actions in the Sensorium beget a sensation of** a mean Colour[87] between their proper Colours; and particularly when

[82] For the importance of Newton's idea that sunlight is not simple, pure and homogeneous, but the most compound of all colours', see Shapiro, 'The Gradual Acceptance of Newton's Theory of Light and Color', pp. 65–6.

[83] For an interpretation of the comb experiments, see Hamou, 'Vision, Color, and Method', pp. 76–80.

[84] See Cohen and Schofield (eds.), *Isaac Newton's Papers*, pp. 116–35, which presents a facsimile of 'Mr. Isaac Newtons [*sic*] Answer to some Considerations upon his Doctrine of Light and Colors', *Philosophical Transactions of the Royal Society* (1672), No. 2, pp. 5084–5103 (note that there are several errors in the original pagination).

[85] For details, see *infra* Pt.IV.4.2, pp. 109–10.

[86] OED *a*: Producing colour or colours (1676–1861), where the first citation is from Newton.

[87] In *Opticks (2)* Newton made a slight emendation to this phrase thus: 'by mixing their Actions in the Sensorium beget a Sensation differing from what either would do apart, that is[,] a Sensation of a mean Colour'; see Newton, *Opticks (C-R)*, p. 159.

> by the concourse and mixtures of all sorts of rays, a white Colour is produced, the white is a mixture of all the Colours which the rays would have apart (Prop. 5, *Lib.* 2.) The rays in that mixture do not lose or alter their several colorifick qualities, but **by all their various kinds of actions mixt in the Sensorium, beget a sensation of** a middling Colour between all their Colours which is whiteness. For whiteness is a mean between all Colours, having it self indifferently to them all....

Bk. I, Pt. II, Prop. VII, Theor. V states: '*All the Colours in the Universe which are made by Light, and depend not on the power of imagination, are either the Colours of homogeneal Lights, or compounded of these and that either accurately or very nearly, according to the Rule of the foregoing Problem.*'[88] Newton then provides a summary of previous proofs that, together, enable him to generalise about all colours in the universe; and Text VI is extracted from this summary.

As for what depends on 'the power of imagination', which he afterwards calls 'the power of phantasy', he mentions, p. 120, colours seen in dreams, by madmen or by those whose eye is struck or pressed by the finger. And towards the conclusion of the book,[89] after describing how to produce a circle of colours by pressing the finger on the bottom *exterior* of the eye, he asks whether or not the same thing happens when rays of light strike the bottom *interior* of the eye (i.e., the retinal coat) and excite vibrations in the retina.

Text VII
Opticks (1), Bk. III,
Queries 11–15, pp. 135–7

> *Qu.* 11. Do not **great Bodies** conserve **their heat** the longest, **their parts heating one another**, and may not **great dense** and fix'd[90] **Bodies**, when heated beyond a certain degree, emit Light so copiously, as by the emission and reaction of **its** [*sic*] Light, and the reflexions and refractions of its [*sic*] rays **within its** [*sic*] **pores** to grow still hotter, till **it** [*sic*] comes to a certain period of **heat**, such as is that of **the Sun**? And are not **the Sun and fix'd Stars** great Earths vehemently hot, **whose heat** is conserved by **the greatness of the Bodies**, and the mutual action and reaction between them, and the Light which they emit, and whose parts are kept from fuming away, not only by their fixity,[91] but also by the vast weight and density of the Atmospheres incumbent upon them, and very strongly compressing them, and condensing the vapours and exhalations which arise from them?
>
> *Qu.* 12. Do not the rays of Light **in falling upon the bottom of the Eye excite vibrations in the *Tunica retina*? Which vibrations, being propagated along the solid**[92] **fibres of the optick Nerves into the Brain, cause the sense of seeing**. For because **dense Bodies** conserve **their heat** a long time, and **the densest Bodies** conserve **their heat** the longest,

[88] The rule referred to in the above proposition is stated in Prop. VI, Prob. II, which includes a geometrical construction based on the analogy between spectral colours and musical pitches ('tones').

[89] See Newton, *Opticks (1)*, Bk. III, Query 16, p. 137,

[90] OED *ppl.a.* 4b: Not easily volatilised; not losing weight under the influence of fire (1641–1830).

[91] OED 1, originally *spec.* in *Physics*: The property of enduring heat without volatilisation or loss of weight (1666–1826).

[92] OED *a.* I. 3, Of material substances: Of a dense consistency (c.1532–1887) = Dense OED 1: Having its constituent particles closely compacted together (1599–1878).

the vibrations of their parts are of a lasting nature, and therefore may be propagated along solid fibres of uniform[93] **dense matter to a great distance, for conveying into the Brain the impressions made upon all the Organs of sense.** For that motion which can continue long in one and the same part of **a Body**, can be propagated a long way from one part to another, supposing **the Body** homogeneal,[94] so that the motion may not be reflected, refracted, interrupted or disordered by any unevenness of **the Body**.

Qu. 13. Do not several sorts of rays make vibrations of several bignesses[95] **excite sensations of** several Colours, much after the manner that the vibrations of the Air, according to their several bignesses **excite sensations of** several sounds? And particularly do not the most refrangible rays **excite** the shortest vibrations for making **a sensation of** deep violet, the least refrangible the largest for making **a sensation of** deep red, and the several intermediate sorts of rays, vibrations of several intermediate bignesses to make **sensations of** the several intermediate Colours?

Qu. 14. May not the harmony and discord of Colours arise from the proportions of **the vibrations propagated through the fibres of the optick Nerves into the Brain**, as the harmony and discord of sounds arises[96] from the proportions of the vibrations of the Air? For some Colours are agreeable, as those of Gold and Indico,[97] and others disagree?

Qu. 15. Are not **the Species of Objects seen with both Eyes united where the optick Nerves meet before they come into the Brain, the fibres on the right side of both Nerves uniting there, and after union going thence into the Brain in the Nerve which is on the right side of the Head, and the fibres on the left side of both Nerves uniting in the same place**, and **after union going into the Brain in the Nerve which is on the left side of the Head, and these two Nerves meeting in the Brain in such a manner that their fibres make but one entire Species or Picture, half of which on the right side of the Sensorium comes from the right side of both Eyes through the right side of both optick Nerves to the place where the Nerves meet, and from thence on the right side of the Head into the Brain, and the other half on the left side of the Sensorium comes in like manner from the left side of both Eyes.** For **the optick Nerves** of such Animals as look the same way with both Eyes (as of Men, Dogs, Sheep, Oxen, *&c.*) **meet before they come into the Brain**, but **the optick Nerves** of such Animals as do not look the same way with both Eyes (as of Fishes and of the Chameleon) **do not meet**, if I am rightly informed.

Of the sixteen queries that conclude Bk. III, those numbered 11–15 constitute Text VI, only some of which mention the *sensorium*. Nevertheless, they are central to our inquiry and will be discussed in Part II.

In *Opticks (2)* the errors in Query 11 identified above with the bracketed word '*sic*' were not corrected. But there were three emendations to Queries 11–14.[98] First, after the final sentence of Query 11, there is a long insertion as follows:

For if Water be made warm in any pellucid Vessel emptied of Air, that Water in the *Vacuum* will bubble and boil as vehemently as it would in the open Air in a Vessel set upon the Fire till it conceives a much greater heat. For the weight of the incumbent Atmosphere keeps

---

[93] OED *a.* I. 2, Of things in respect of their own qualities or constitution: Having or presenting the same form or appearance (1550–1884).

[94] OED A. *a.*, now rare = Homogeneous *a.* 2, Of a thing in respect of its constitution: Of uniform nature throughout (1695–1859).

[95] I.e., vibration lengths, periods or intervals, a meaning not included in OED under the term 'bigness'.

[96] I.e., arise; corrected in *Opticks (2)*.

[97] I.e., indigo; corrected in *Opticks (2)*.

[98] See Koyré, 'Les Queries de l'Optique', p. 19.

down the Vapours and hinders the Water from boiling, until it grow much hotter than is requisite to make it boil *in vacuo*. Also a mixture of Tin and Lead being put upon a red hot Iron *in vacuo* emits a Fume and Flame, but the same Mixture in the open Air, by reason of the incumbent Atmosphere, does not so much as emit any Fume which can be perceived by Sight. In like manner the great weight of the Atmosphere which lies upon the Globe of the Sun may hinder Bodies there from rising up and going away from the Sun in the form of Vapours and Fumes, unless by means of a far greater heat than that which on the Surface of our Earth would very easily turn them into Vapours and Fumes. And the same great weight may condense those Vapours and Exhalations as soon as they shall at any time begin to ascend from the Sun, and make them presently fall back again into him, and by that action increase his Heat much after the manner that in our Earth the Air increases the Heat of a culinary Fire. And the same weight may hinder the Globe of the Sun from being diminish'd, unless by the Emission of Light, and a very small quantity of Vapours and Exhalations.[99]

Second, there is an addition to the first sentence of Query 13: 'Do not several sorts of Rays make Vibrations of several bignesses, which according to their bignesses excite Sensations of several Colours,....'[100] Third, the final sentence of Query 14 is emended thus: 'For some Colours, if they be view'd together, are agreeable to one another, as....'[101]

Text VIII
Optice,
Query 20 (pp. 307–15) p. 314–5

Qui fit, ut ***Natura nihil Agat frustra****? & Unde orta est eximia illa Mundi universi Species & Pulchritudo? ... Qui fit, ut Corpora Animalium tam exquisita sint Arte atq; Consilio*[102] *fabricata? & Quos ad fines conformatæ sunt diversæ ipsorum Partes? Fierine potuit, ut Oculus sine scientia Optices fuerit constructus? aut Auris, sine Intelligentia Sonorum? Qui fit, ut Motus Corporis obsequantur* ***Imperio Voluntatis****? & Unde est Instinctus ille quem vocant, in Animalibus?* ***Annon Sensorium Animalium, est locus cui Substantia sentiens adest, & in quem sensibiles rerum species per nervos & cerebrum deferuntur, ut ibi præsente sentiri possint?*** Atq; his quidem rite expeditis, *Annon ex phænomenis constat, esse* ***Entem Incorporeum, Viventem, Intelligentem, Omnipræsentem, qui in Spatio infinito, tanquam Sensorio suo, res Ipsas intime cernat, penitusq; perspiciat, totasq; intra se præsens præsentes complectatur; quarum quidem rerum Id quod in nobis sentit & cogitat, Imagines tantum ad se per Organa Sensuum delatas, in Sensoriolo suo percipit & contuetur?*** Utiq; si verus omnis in hac Philosophia factus progressus, non quidem statim nos ducit ad *Causæ primæ* cognitionem; at certe propius propiusq; nos ad eam perpetuo adducit, eaque re permagni est æstimandus.

How does it happen that **nature does nothing in vain**?[103] And where did the wonderful splendour and beauty of the world come from? ... How does it happen that the bodies of

[99] See Newton, *Opticks (C-R)*, pp. 344–5.

[100] Ibid., p. 345.

[101] Ibid., p. 346.

[102] The conjunction 'atque' denotes a closer internal connection than is implied by 'et' and gives prominence to what follows, i.e., '*Consilio*'.

[103] This maxim, which has teleological import, is repeated by Newton in Text XI below. It was Aristotle's principle of economy; see Van Helden, *Measuring the Universe*, p. 46. And it was also an anatomical maxim of Galen of Pergamum; see Woollam, 'Concepts of the Brain', p. 18 where

> animals were fashioned with such exquisite skill and purpose? For what ends were the different bodily parts shaped? Was it possible that, without a knowledge of optics, an eye was made? Or an ear, without an understanding of sounds? How does it happen that **the motions of a body obey the dictates of the will**? And where does animal instinct[104] come from? **Is not the sensory of animals a place where a sentient[105] substance is present and it is into this place that the sensible species of things are transferred through the nerves and the brain so that, when said substance is present, they may be perceived**? When these things are duly performed, is it not evident from the phenomena that there exists **an incorporeal, living, intelligent[106] and ever-present being who in infinite space, as if** [in] **his sensory, intimately sees and thoroughly perceives things themselves and, himself present, comprehends[107] the things present**? **Does not that which in us senses and reflects**[,] **perceive and survey the images of these things, which are conveyed to it by the sense organs, into its little sensory**? Hence, if all progress achieved in this philosophy is true, this does not lead us immediately to acquaintance with the first cause but certainly leads us ever closer to it, which is something of very great importance. (*Translation by David Fairservice*)

In the 1706 *Optice* seven new queries were added to the original sixteen and numbered 17–23. The extract from Query 20 reproduces part only of Newton's teleological argument from the Latin of Samuel Clarke. In *Opticks (2)*, Queries 17–23 were renumbered and given an English dress by Newton himself; and at that time, Query 20 became Query 28. Consequently, Newton's English version of the extract from Query 20 is provided for ease of comparison with the extract from Query 28, for which see Text XI below. In this latter text Newton translated the term '*Substantia sentiens*' not as sentient substance, but as 'sensitive substance'; and, as will appear in the note to Text XI, the meanings of the terms 'sentient' (see n.105 below) and 'sensitive' (see p. 24 n.136 below) are not identical.

Text IX
Optice,
Query 23 (pp. 322–48) pp. 346–7

> Tam miram **uniformitatem in Planetarum Systemate**, necessario fatendum est **Intelligentia & Consilio** fuisse effectam. Idemq; dici possit de **uniformitate** illa, quæ est **in Corporibus *Animalium***. ... Deinde Partes illæ Corporis, tam exquisita Arte atq[108];

Galen is quoted as follows: 'You would think that Nature, who apparently does nothing in vain, would have made this aperture to no purpose; but he who dissects carelessly, commits crimes of this kind not only against anatomy itself but against its natural maxims.'

[104] OED *sb.* 1, obs.: Impulse (1423–1730); also OED *sb.* 3, *spec.*: An innate propensity that manifests itself in acts which appear to be rational but are performed without conscious design or intentional adaptation of means to ends (1596–1877).

[105] OED *a.* A. 1 that feels or is capable of feeling; having the power or function of sensation or of perception by the senses (1635–1865); hence, a sentient substance is a substance that feels.

[106] OED B *sb.* 1: An intelligent or rational being = Intelligence 4: An impersonation of intelligence, esp. applied to one that is, or may be an incorporeal spirit (1378–1837).

[107] OED *v.* II. 4 *trans.*: To grasp (with the mind), understand, 'take in' (1340–1886); also OED *v.* III. 10b, obs.: To contain, 'take in' (1631–1806).

[108] The conjunction 'atque' denotes a closer internal connection than is implied by 'et' and gives prominence to what follows, i.e., '*Consilio*'.

Consilio fabricatæ, Oculi, Aures, Cerebrum, Musculi, Glandes, Cor, Pulmones, Diaphragma, Larynx....; Sensus & Motus Organa, Instinctusq; in Animalibus brutis & insectis; horum sane omnium conformatio prima, nulli rei tribui potest, nisi **Intelligentiæ & Sapientiæ Entis Potentis semperq; Viventis; quod sit ubiq; scilicet præsens, possitq; Voluntate sua corpora omnia in infinito suo *Sensorio* movere**, adeoq; cunctas Mundi universi partes ad arbitrium suum fingere & refingere, **multo magis quam Anima nostra, quæ est in Nobis Imago Dei**, **voluntate sua** ad corporis nostri membra movenda valet.

It must be conceded that such wonderful **uniformity**[109] **in the planetary system** was the effect of **intelligence**[110] **and choice.**[111] The same may be said of **the uniformity** to be found **in the bodies of animals**. ... Then the parts of the body, framed with such exquisite skill and purpose, eyes, ears, brain, muscles, glands, heart, lungs, diaphragm, larynx...; the organs of sense and motion, and instinct[112] in brute animals and insects; the first shaping of all these cannot be attributed to anything but to **the intelligence and wisdom**[113] **of a powerful, ever-living, being, which is certainly present everywhere and is able by his will to move all bodies in his infinite sensory** and hence to form and re-form all the parts of the whole world **at will**[114] and is able, far more than **our soul, which in us is the image of God,** is able **by our will** to move the members of our body. (*Translation by David Fairservice*)

Text IX is extracted from another argument from design in Query 23, which was renumbered as Query 31 in *Opticks (2).* But Newton himself, who provided the English version of Query 31, made a number of emendations (additions and alterations), as will be evident by comparing Text IX above with Text XII below. Of particular importance is Newton's deletion from Text IX of the words 'Anima nostra, quæ est in Nobis Imago Dei ('our soul, which in us is the image of God in us').

It should also be noted that in the original issue of *Optice*, of which only five copies survive, there was a different version of part of the second passage in Latin reproduced above in bold face, namely, that which begins '*Annon ex phænomenis constat*' and ends with the words '*in Sensoriolo suo percipit & contuetur*'. In the original version, the passage read as follows:

Annon Spatium Universum, **Sensorium** est **Entis Incorporei, Viventis, et Intelligentis**; quod Res Ipsas cernat et complectatur intimas, totasque penitus et in se præsentes perspiciat; quarum id quidem, **quod in Nobis sentit et cogitat, Imagines tantum in Cerebro contuetur**?

Is not the whole of Space **the Sensorium of a Being Incorporeal, living, and intelligent**; in that he sees distinctly and closely comprehends[115] the most inward things themselves,

---

[109] OED 2: Regularity in the action, here, of providence (1577–1875); also OED 3: Similarity of design (1625–1842).

[110] OED *sb*. 1: The faculty of understanding; intellect (1390–1830).

[111] OED *sb*. 2: The faculty of choosing (1300–1842).

[112] OED 1, obs.: Impulse (1412–1730); also OED 3 *spec*.: An innate propensity manifesting itself in acts which appear to be rational, but are performed without conscious design or intentional adaptation of means to ends (1596–1877),

[113] OED 1a: Soundness of judgment in the choice of means and ends (c.1000–1875).

[114] OED III. Special uses, 15: According to volition or choice (1300–1802).

[115] OED *v*. II. 4 *trans*.: To grasp (with the mind), understand, 'take in' (1340–1886); also OED *v*. III. 10b, obs.: To contain, 'take in' (1631–1806)

> and observes them wholly and thoroughly by their being present **in him**; of which [things], certainly, **that which in us senses and reflects, surveys merely the Images in the Brain**? (*Translation by F. E. L. Priestley*)[116]

An article by Alexandre Koyré and Bernard Cohen, based on the discovery of the original version,[117] was followed by a critique, as well as by a number of attempts to explain, theologically, why Newton may have cancelled the passage.[118] Note, however, that a non-theological reason could also be offered. For the Latin passage from the first issue seems to repeat, with slight variation, the passage in the previous Text VIII beginning with the words '*esse Entem*' and ending with '*percipit & contuetur*'. To my knowledge, this fact has not been noticed; but it could have led Newton to order the page or signature cancelled and replaced with another. See also *infra* Pt. IV.4.2, p. 112, for another reason why he might have cancelled the passage.

Text X
Opticks (2)
Queries 23 and 24, p. 328

> *Qu.* 23. Is not Vision perform'd chiefly by the Vibrations of this [ætherial] Medium, **excited in the bottom of the Eye** by the Rays of Light, **and propagated through the solid,**[119] **pellucid**[120] **and uniform**[121] **Capillamenta**[122] **of the optick Nerves into the place of Sensation**?[123] And is not Hearing perform'd by the Vibrations either of this or some other Medium, **excited in the auditory Nerves** by the Tremors of the Air, **and propagated through the solid, pellucid and uniform Capillamenta of those Nerves into the place of Sensation**? And so of the other Senses.
>
> *Qu.* 24. Is not Animal Motion perform'd by the Vibrations of this Medium, **excited in the Brain by the power of the Will,**[124] and propagated from thence **through the solid, pellucid and uniform Capillamenta of the Nerves into the Muscles, for contracting and dilating them**? I suppose that **the Capillamenta of the Nerves are each of them solid and uniform**, that the vibrating Motion of the Ætherial Medium may be propagated **along them** from one end to the other uniformly, and without interruption: For Obstructions in the Nerves create Palsies.[125] And that they may be sufficiently uniform, I suppose them to be

---

[116] See Priestley, 'The Leibniz-Clarke Controversy', p. 37.

[117] See Koyré and Cohen, 'The Case of the Missing *Tanquam*'.

[118] For the critique, see Priestley, 'The Leibniz-Clarke Controversy', pp. 37–41. For some theological explanations, see Westfall, *Never at Rest*, pp. 647–8, and Connolly, 'Newton and God's Sensorium', p. 195.

[119] OED *a.* I. 3, Of Material substances: Of a dense consistency (c.1532–1830) = Dense OED 1: Having its constituent particles closely compacted together (1599–1878).

[120] OED *a.* 1: Having the property of transmitting, or allowing the passage of light; translucent, transparent, clear (1619–1863).

[121] OED *a.* I. 2, Of things in respect of their own qualities or constitution: Having or presenting the same form or appearance (1550–1884).

[122] I.e., hair-like fibres or filaments, the ultimate ramifications in living matter; see also *infra* Pt.II.2.1, pp. 35, 37–38.

[123] I.e., 'the place' = the *sensorium*.

[124] OED *sb.*[1] 6: The power of choice in regard to action (c.888–1880).

[125] OED *sb.* 1: Disease of the nervous system, characterised by impairment or suspension of voluntary muscular action or sensation and, in some forms, by involuntary tremors of the limbs (13..–1860).

> pellucid when view'd singly, tho' the Reflexions **in their cylindrical Surfaces** may make **the whole Nerve (composed of many Capillamenta**) appear opake[126] and white. For opacity[127] arises from reflecting Surfaces, such as may disturb and interrupt the Motions of this Medium.

In *Opticks (2)*, Newton inserted seven new queries between Queries 1–16 of the first 1704 edition and Queries 17–23 added to the 1706 Latin translation. The newly inserted queries were numbered 17–24; and those added in 1706 were re-numbered as Queries 24–31. In Queries 18–20 Newton introduced an 'Æthereal Medium' for vision, at the same time acknowledging that each sensory mode may have a different kind of medium. In Text X, which is a highly condensed version of Text II, Query 23 reiterates the possibility that different media are required for each sensory mode; and Query 24 merely mentions that 'Animal Motion' is performed by the vibrations of 'this' (presumably, ætherial) medium.

In the latter query, which concerns voluntary actions, Newton includes a brief observation that obstructions in the nerves create (involuntary) palsies. This disease is mentioned several times, but not elaborated on, in Willis's 1667 book, *Pathologiæ cerebri, et nervosi,* in which he stated that various kinds of convulsive motions 'never happen but in a living Body, where the Nervous parts are blown up, and grow turgid with the animal spirit'.[128] And as previously indicated (see comment to Texts II and IV above), Newton owned a 1668 edition of this book. But in an entire chapter on palsy in his 1672 book, *De anima brutorum,* Willis set out three ways in which obstructions occur in the nerves, as well as in other parts of the nervous system.[129]

Why Newton returned to an ætherial medium is puzzling, especially because in his 'Conclusio', a suppressed manuscript conclusion to the second, 1713 edition of *Principia mathematica,* written sometime in the spring of 1687, he relied only on the chemical powers of attraction and repulsion to claim that

> ...in the same way that in air that vibratory motion of which sound consists ... [so] also those motions by which the Tunica Retiformica is agitated by light ... **can be propagated to the sensorium through the solid and continuous capillamenta of the nerves**. And on the other hand, by a similar motion **propagated from the sensorium through the solid capillamenta of the nerves, a certain substance in the muscles can be agitated and by that agitation dilated so as to contract the muscles and move the limbs. For the animal spirits** (which they[130] feign) **can hardly be propagated easily, swiftly and copiously enough through the compact substance of the nerves to swell the muscles**. And those who hold up heavy weights are heated to sweating even without moving their limbs. **That heat does not arise from the inrush of animal spirits into muscles which were already full before, but from a certain agitation of the part[icle]s of the body by which the muscles are kept distended, and when this ceases they grow flaccid. Furthermore through the slow and continued motion of heat the particles of bodies can gradually**

[126] I.e., opaque; see OED *a.* 2: Impermeable to light, not transmitting light; hence, impenetrable to sight (1641–1860).

[127] OED 2: The quality or condition of being impervious to light (1634–1885).

[128] Willis, *An Essay on the Pathology of the Brain* [1665 tr. 1684], p. 3.

[129] See Willis, *Two Discourses* [1672 tr. 1684], Ch. IX, pp. 161–78.

[130] Possibly, the Cartesians.

> **change their arrangement and coalesce in new ways and by the attractive forces of contiguous particles (which are stronger than expulsive ones) come together more densely.**[131]

Text XI
Opticks (2),
Query 28 (pp. 335–45[132]) pp. 344–5

> Whence is it that **Nature doth nothing in vain**[133]; and whence arises all that Order and Beauty which we see in the World? ... How came the Bodies of Animals to be contrived with so much Art, and for what ends were their several Parts? Was the Eye contrived without Skill in Opticks, and the Ear without Knowledge of Sounds? How do the Motions of the Body follow from **the Will,**[134] and whence is the Instinct[135] in Animals? Is not **the Sensory of Animals that place to which the sensitive**[136] **Substance is present, and into which the sensible Species of Things are carried through the Nerves and Brain, that there they may be perceived by their immediate presence to that Substance**? And these things being rightly dispatch'd, does it not appear from Phænomena that there is **a Being incorporeal, living, intelligent,**[137] **omnipresent, who in infinite Space, as it were in his Sensory, sees the things themselves intimately and thoroughly perceives them, and comprehends**[138] **them wholly by their immediate presence to himself**: Of which things the Images only[,] carried **through the Organs of Sense into our little Sensoriums, are there seen and beheld by that which in us perceives and thinks.**[139] And though every true Step made in this Philosophy brings us not immediately to the Knowledge of the first Cause, yet it brings us nearer to it, and on that account is to be highly valued.

Text XI presents Newton's own translation of, as well as emendations to that part of a teleological argument extracted in the foregoing Text VIII from Query 20 of

---

[131] See Newton, 'Conclusio', pp. 345–6 (bold-face is mine); see also the editors' preface to the manuscript, pp. 320–1.

[132] For emendations to the whole of Query 28, see Koyré, 'Les Queries de l'Optique', pp. 22–3.

[133] This maxim, which has teleological import, is repeated by Newton in Text VIII above. It was Aristotle's principle of economy; see Van Helden, *Measuring the Universe*, p. 46. And it also was a maxim of the anatomist, Galen of Pergamum; see Woollam, 'Concepts of the Brain', p. 18, where Galen is quoted as follows: 'You would think that Nature, who apparently does nothing in vain, would have made this aperture to no purpose; but he who dissects carelessly, commits crimes of this kind not only against anatomy itself but against its natural maxim.'

[134] OED *sb.*[1] 5b,?obs.: Intention, purpose (1300–1712).

[135] OED *sb.* 1, obs.: Impulse (1412–1730); also OED *sb.* 3. *spec.*: An innate propensity that manifests itself in acts which appear to be rational but are performed without conscious design or intentional adaptation of means to ends (1596–1877).

[136] OED *a.* A.2: Of living beings: Endowed with the faculty of sensation (1555–1796), Formerly often, obs.: Having sense or perception, but not reason (1555–1755).

[137] OED B. *sb.* 1: An intelligent or rational being = Intelligence 4: An impersonation of intelligence, esp. applied to one that is, or may be an incorporeal spirit (1378–1837).

[138] OED *v.* II. 4 *trans.*: To grasp (with the mind), understand, 'take in' (1340–1886); also OED *v.* III 10b, obs.: To contain, 'take in' (1631–1806).

[139] OED *v.*[2] B.II. 3 *intrans.*: To exercise the mind, esp. the understanding, in any active way; to make a train of ideas pass through the mind; to meditate, cogitate (c.1000–1864). According to OED, this is the 'most general verb to express internal mental activity, excluding mere perception of external activity or passive reception of ideas'.

Optice. Regarding his substitution of the term 'sensitive substance' for the Latin term '*Substantia sentiens*', see also the comment and note to Text VIII above.

Text XII
Opticks (2),
Query 31 (pp. 349–82[140]) pp. 378–9[141]

> Such a wonderful **Uniformity**[142] **in the Planetary System** must be allowed the Effect of **Choice.**[143] And so must **the Uniformity in the Body of Animals**.... Also the first Contrivance of those very artificial Parts of Animals, the Eyes, Ears, Brain, Muscles, Heart, Lungs, Midriff, Glands, Larynx.... and other Organs of Sense and Motion; and the Instinct[144] of Brutes and Insects, can be the effect of nothing else than the **Wisdom**[145] **and Skill of a powerful ever-living Agent,**[146] **who being in all Places, is more able by his Will to move the Bodies within his boundless uniform**[147] **Sensorium,** and thereby to form and reform the Parts of the Universe, than we are **by our Will** to move the Parts of our own Bodies. And yet we are not to consider the World as the Body of God, or the several Parts thereof, as the Parts of God. **He is an uniform**[148] **Being**, void of Organs, Members or Parts, and they are his Creatures subordinate to him, and subservient to **his Will; and he is no more the Soul of them, than the Soul of Man is the Soul of the Species of Things carried through the Organs of Sense into the place**[149] **of its Sensation, where it perceives them by means of its immediate Presence, without the Intervention of any** [mediator or] **third thing. The Organs of Sense are not for enabling the Soul to perceive the Species of Things in its Sensorium, but only for conveying them thither; and God has no need of such Organs, he being every where present to the Things themselves.**

Text XII, extracted from the concluding Query 31, presents a revised version of part of the argument from design in *Optice*, Query 23, for which see Text IX. A comparison of these texts will indicate that the earlier Text IX appears here in an

---

[140] For emendations to the whole of Query 31, see Koyré, 'Les Queries de l'Optique', pp. 27–8.

[141] Part of this extract from Qu. 31 may have been written in answer to Jonathan Edwards, who in 1714 inferred from Newton's claims, first, that 'the *Infinitum Sensorium* ... belongs to God, and *in which he moves all Bodies as he pleases, as the Soul that is in Man moves the Members of his Body*' and, second, that '*Organs of Sense and Motion* belong to God, as well as to Man' and, hence, 'God *to have a Body*, which is Unphilosophical..., as well as Untheological'; see Edwards, 'Postscript', p. 38.

[142] OED 2: Regularity in the action, here, of providence (1577–1875); also OED 3: Similarity of design (1625–1842).

[143] OED *sb.* 2: The faculty of choosing (1300–1842).

[144] OED 1, obs.: Impulse; also OED 3 *spec.*: An innate propensity manifesting itself in acts which appear to be rational, but are performed without conscious design or intentional adaptation of means to ends (1596–1873).

[145] OED 1a: Soundness of judgment in the choice of means and ends (c.1000–1875).

[146] OED B. *sb.* 1: One who (or that which) exerts power as distinguished from the patient and also from the instrument (1600–1870).

[147] OED *a.* I. 1: Of one form, exhibiting no difference, diversity or variation (1540–1891).

[148] OED *a.* I. 1 *absol.*: 'The Uni-form which gives all forms their Being' (1606). Note that in this definition the 'Uni-form' is one in number and, hence an arithmetical unit. See further *infra* Conclusion, pp. 132–3.

[149] I.e., 'the place' = the *sensorium*.

emended state that includes both additions and deletions. Newton, however, has created a problem in the convoluted sentence that begins 'He is an uniform being' and ends 'without the Intervention of any third thing' (i.e., without a 'Substantia sentiens' or 'sensitive substance'). For in that sentence Newton uses the pronouns 'he', 'his' or 'him' to refer to God, whereas he uses the terms 'its' and 'it' to refer to the 'Soul of Man', which he claims, does not need 'any third thing'. This claim contradicts not only the foregoing Texts VIII and XI, but also the concluding sentence in the above text, which states that God perceives things directly, whereas humans perceive things indirectly, that is, by a mediator or 'third' thing.

Part II

II
The Human *Sensorium* in Context

2.1 Newton on the Sensory-Motor System

In the foregoing part I presented Newton's texts as they relate to the human, as well as the divine *sensorium*. The purpose of this and the next section is to discover what can be learned from those about the human *sensorium*. Although this usage of the term '*sensorium*' is found most frequently in the texts concerning the visual sensory system, yet I shall begin with Text III, in which there is no mention of a human *sensorium*. Nevertheless, an analysis of that text, as well as evidence and inferences drawn from it, will help to identify problems that require solution during the course of this inquiry.

The source for Text III is *Opticks (1)*, Book I, Part I, Axiom VII, which states:

> Wherever the Rays which come from all the Points of any Object meet again in so many Points after they have been made to converge by Reflexion and Refraction, there they will make a Picture of the Object upon any white Body on which they fall.'[1]

Two paragraphs of comment follow this Axiom, in the first of which Newton describes 'a Vulgar Experiment of casting the Species of Objects from abroad upon a Wall or Sheet of white Paper in a dark Room'. This experiment involves an external object, a lens placed at a hole in the glass of a window 'shut in a dark Chamber' and a piece of white paper held so that the light falls upon it. He also points out that the 'Picture' on the paper reproduces the external object in shape and colour.

The second paragraph, from which Text III is extracted, commences with the words 'In like manner', a hint that just as a 'Picture' is formed in the interior of a 'dark Chamber', so too a 'Picture' is formed in the interior of the eye, with this difference: that in the latter, 'the Picture shall be inverted'.[2] Accordingly (see *supra* Pt. I.1.2, Text III, Fig. I.1), Newton conceives the pupil (*mk*) as the hole in a shut window; the coat of the cornea (*EFG*) as the glass in the window; the crystalline

[1] Newton, *Opticks (1)*, Axiom VII, pp. 9–11, p. 9.

[2] Ibid., p. 10.

J. C. Kassler, *Newton's* Sensorium*: Anatomy of a Concept*, Archimedes 53,
https://doi.org/10.1007/978-3-319-72053-1_2

humour (*AB*) as a lens placed at the hole; and the coat of the retina ('*Tunica Retina*') that covers 'the bottom of the Eye' as the white paper. For, as he indicates, 'Anatomists', by taking off the outermost covering of the brain, are then able to see through the thinner coats of the retina. And what they see are 'Pictures of Objects lively painted thereon.'

The brief description in Text III concerns the anatomical pathway for *optical* processes. But it also contains a hint concerning the anatomical pathway for *visual* processes, namely, that vision itself is caused only when the 'Pictures' formed 'on' the coat of the retina are transmitted 'along' the fibres of the optic nerves 'into' the brain. But there is no detail concerning the mode in which 'Pictures' are transmitted from the retina to the brain. Instead, the principal aim of this text is to explain image formation 'on' the retinal coat. And to do this, Newton conceives the eye as an optical instrument based on the 'Vulgar' experiment in a 'dark Room or Chamber', that is to say, in a *camera obscura.*[3]

This experiment became commonplace ('Vulgar') after Johannes Kepler used it to demonstrate how the eye formed an inverted image on the retina, an achievement that has been described as 'the first major modern physiological discovery'.[4] Prior to this demonstration, it was supposed that the crystalline humour and the retina were 'multiplying' (magnifying) glasses, the former serving as the receptive part of the eye. Kepler, however, explained precisely how rays of light coming from a given object are refracted by the coats or surfaces of the cornea and by the crystalline humour so as to delineate a small inverted image of the object on the surface of the retina, thereby demonstrating that the retina, not the crystalline humour, is the receptive part of the eye.[5] Like some of his predecessors, he used the experiment in the *camera obscura* to model the interior of the eye as an optical instrument with a spherical lens. But in executing the optical analysis correctly, he demonstrated that the crystalline humour functions as a refractor and not, as previously supposed, a receptor.[6]

After Kepler, other natural philosophers improved on his account.[7] For example, in 1619 Christoph Scheiner added more lenses,[8] and in 1637 René Descartes

---

[3] Newton, *Opticks (1)*, p. 10. For other references to a 'dark Chamber' or 'dark Room', see ibid., Bk. I, Pt.I, pp. 9, 15, 18, 28, 30, 34, 36, 39, 41, 47, 50, 55; Pt.II, pp. 81, 83–4, 98, 105, 121; Bk. II, Pt. I, p. 15, Pt.III, pp. 52, 66, Pt. IV, pp. 87–8; Bk. III, pp. 113, '112' [=120], 124, 129. For earlier references, see Cohen and Schofield (eds.) *Isaac Newton's Papers & Letters,* pp. 48, 56–7, 59, 91, 136–8, 145, 156, 160, 170, 172, 192–3, 195–8, 207, 234, *et passim*. It is not clear from these references whether Newton was using a room converted into a dark chamber or whether he was using a portable box-shaped structure of room size.

[4] Crombie, *Science, Optics and Music*, p. 170. The second major modern physiological discovery was made by William Harvey concerning the circulation of the blood; for details of publication, see the 'Editor's Postscript' at the conclusion of Harvey, *The Anatomical Exercises*.

[5] See Koelbing, 'Ocular Physiology', pp. 219–24.

[6] See ibid., p. 226.

[7] For a detailed treatment of the problem of optical imagery from Kepler to the early part of the eighteenth century, see Shapiro, 'The Optical Lectures'.

[8] See Crombie, *Science, Optics and Music*, p. 232 and fig. 36, p. 233.

corrected Kepler's optical analysis by making use of his new sine law of refraction by treating the shape of the crystalline humour as a convex, not a spherical lens.[9] Text III, however, retains two refracting surfaces—the cornea and crystalline humour—and, hence, represents a highly condensed version of Kepler's functional anatomy as presented in two different books—one, *Ad Vitellionem paralipomena quibus astronomiæ pars optica traditur* (1604), the other, *Dioptrice* (1611).[10] Since Barrow owned copies of the first editions of both books, Newton would have had access to them.[11] But he later obtained for his own library a copy of the second book in its third edition published at Amsterdam in 1682.[12]

It is important to note, therefore, that Kepler drew attention to a physiological problem that falls outside the domain of optical processes. For in his 1604 book, he called on natural philosophers

> ...to discuss the way in which this image or picture (*pictura*) is put together by the spiritual principles of vision residing in the retina and in the [optic] nerves, and whether it [the *pictura*] is made to appear before the soul or tribunal of the faculty of vision by a spirit within the cerebral cavities, or [whether] the faculty of vision, like a magistrate sent by the soul, goes out from the council chamber of the brain to meet this image in the optic nerves and retina, as it were descending to a lower court. [*Translation by A. C. Crombie* (additions in square brackets mine)][13]

Then, in 1611 he himself took the part of a natural philosopher in order to offer a brief explanation concerning what happens *after* the picture is painted 'on' the coat of the retina, for

> ...this picture does not complete the act of vision until the image (*species*) so received by the retina passes through the continuity of the spirits to the brain, and is there delivered to the threshold of the faculty of the soul. This happens as follows: all external sense is accomplished by a reception and impression (*impressio*), that is by an affection (*passio*), when the image (*species*) of an external thing is impressed on that which feels, and this affection is called sensation (*sensio*). But inside within the brain is something, whatever it may be, which is called the *sensus communis*, on which is impressed the image (*species*), effecting the visual affection, that is painted by the light from the visible thing. Thus these images fall on the instrument of vision [the retina] outside the seat of the *sensus communis*; then by means of an immaterial image (*species immateriata*) descending from the instrument affected or painted and carried to the threshold of the *sensus communis* they [the *species immateriata*] are impressed on the *sensus communis*. But this impression is hidden

---

[9] Ibid., pp. 232–6, *et passim*. For Descartes's indebtedness to Kepler, see Riba, 'Cartesian Optics', pp. 45, 52.

[10] For a brief description of the astronomical content of the two books, see Gingerich, 'Johannes Kepler', pp. 298–9. Note that in the latter book, Kepler included a description of the optics of spherical lenses, which Riba, 'Cartesian Optics', p. 58, described as a direct descendant of the modern theory of lenses.

[11] See Feingold, 'Isaac Barrow's Library', (Nos. 559 and 560), p. 336. According to Shapiro, 'The Spectre of Newton's "Spectrum"', p. 186, Newton did not read the first book before 1672.

[12] See Harrison, *The Library of Isaac Newton*, (No. 651), p. 147.

[13] Crombie, *Science, Optics and Music*, pp. 297–8. In the last part of this long sentence, Kepler alludes to Plato's *synaugeia* theory of sensory perception, according to which sensation is not a one-directional process; see p. 51 and n.101 below.

from our understanding. [*Translation by A. C. Crombie* (additions in square brackets mine)][14]

There are four points to note concerning the two passages quoted above. First, Kepler made an unexplained distinction between the image (*pictura/species*) impressed on the coat of the retina and the image (*species immateriata*) transmitted from the retina and impressed on 'something' called *sensus communis*.[15] Newton also makes a similar distinction in his use of the term 'image' and its cognates. In Texts I, III, V and VIII the terms 'species' and 'pictures' are used to mean the *outward* appearance of something as constituting the immediate object of vision.[16] But in Texts VII, VIII and XI the terms 'species', 'images' and *sensibiles rerum species*' (sensible species of things) are used to mean an emission from outward things that form the direct objects of sensory perception.[17] Note, therefore, that the two meanings relate to two different processes, optical and visual. Consequently, in the rest of this inquiry, I shall indicate this distinction as image[a] and image[b].

Second, in both quotations from Kepler there are a number of unexplained entities—'spiritual principles' of vision in the retina, optic nerves and 'cerebral cavities'; a 'faculty' inside the brain called *sensus communis*; and the soul that is the 'tribunal' of this faculty. Newton's texts also contain a number of unexplained entities, including 'ætherial vital spirits' in the circulatory system and 'ætherial animal spirits' in the nervous system[18]; a 'sentient substance' ('Subtantia sentiens') or 'sensitive substance' in the *sensorium*[19]; and a 'soul' in the brain.[20] In Newton's case, however, all these names are merely terms of convenience for unknown causes, as will become evident in the next section.

Third, even though Kepler's emphasis was physiological, not anatomical, he provided no information concerning (1) the mode of transmission of the *species immateriata*; (2) their anatomical pathway from the retina to the 'threshold' of the soul; and (3) the specific name, 'seat' or place in the anatomical space of brain in which the common-sense faculty is impressed. Consequently, if Newton were to improve on the two quotations from Kepler (and I am not suggesting that this was his aim), then he would need to provide the necessary information.

[14] Ibid., pp. 238–9. Note that Crombie, pp. 284–322, provided what is probably the first English translation of the whole of Chapter V, Section 2 of *Ad Vitellionem paralipomena*, together with the introduction to this chapter, the introductions to Sections 1 and 3, and the Propositions in Section 3 demonstrating the main conclusions about the formation of the retinal image described in Section 2.

[15] The notion of a *sensus communis* is traceable to Aristotle, who distinguished between special objects, each of which are definitive of one of the five external senses, and 'common-sense objects', which are available only to a common sense; see Hett (tr.), *Aristotle On the Soul*, III.i, pp. 143–5, Lawson-Tancred (tr.), *Aristotle De Anima*, III.i, pp. 189–91. See *infra* Pt. III.3.3, p. 89.

[16] See OED I.3., obs. (1588–1690).

[17] See OED I. *Metaphysics*: 5.a and b, obs. (1614–1707 and 1621–1726).

[18] See *supra* Pt. I.1.2, Text II and comment to Text VIII.

[19] See *supra* Pt. I.1.2, Texts VIII and XI.

[20] See *supra* Pt. I.1.2, Texts II and X.

Fourth and finally, in the texts of both Kepler and Newton, there is a reliance on a container/contained model. Indeed, a careful reader of their texts will have noticed such important little words as 'in/into/within', 'along/through', 'on/upon', which are symptomatic of this model. Unlike Kepler, who revealed little about his containers, Newton is more or less forthcoming about them, for in the analogy of nature presented in Text VII (Ques. 11, 12), we learn three important things about those containers.

Firstly, the size of containers ranges from extremely large—for example, in the solar system, the sun and fixed stars—to extremely minute—for example, in the nervous system, the hypothetical fibres.[21] Secondly, regardless of size, all containers are dense bodies with pores and, thus, have a compound structure; that is to say, they are structured composites consisting of a hierarchy of different sized particles and, hence, different sized pores. Accordingly, since containers are not hollow but have channels in which something is contained, Newton's theory of compound matter may be classed as a porosity theory.[22] Finally, in addition to the eye and some of its parts (Kepler's specialty), Newton identifies a number of other containers as follows: the muscles, including the heart; the nerves and their parts (the fibres); the brain and one of its parts (the *sensorium*).

In 1664 Thomas Willis defined such containers as structures that are 'gifted with animal Spirit', on which 'Motion and Sense necessarily and immediately depend ... for the performing either one only, or both together in the whole Body'.[23] Consequently, if the containers of the nervous system contain animal spirits, then for the performance of sensory-motor functions, there must be anatomical pathways for the transmission of these entities.

In Texts II, VII and IX, Newton merely outlines this pathway insofar as it concerns parts of the visual sensory system, including in Text X the fibres under 'cylindrical surfaces' of the optic nerves of each eye. In Text II he also indicates that the ends of the fibres, which exit behind the retina of each eye, are refracting surfaces ('superficies'), as well as the place where light rays excite vibrations that are transmitted through the optic nerves to the *sensorium*. And in one important but convoluted sentence in Text VII (Qu. 15), he provides more detail as follows: after exiting behind the retina, the two optic nerves 'meet' or 'unite' before going into the brain, after which one nerve goes to the exterior surface of the right side of the *sensorium* and the other, to the exterior surface of its left side. We may infer, therefore, first, that the anatomical substrate called 'the *sensorium*' is the place in the brain where the optic nerves (and their fibres) terminate, and second, that the optic nerves themselves form an X-shaped structure.

---

[21] See Edwin Clarke, 'The Doctrine of the Hollow Nerve', p. 124; see further pp. 37–8 below.

[22] See Shapiro, *Fits, Passions, and Paroxysms*, pp. 86–8, *et passim*, for what he called Newton's 'compositional theory of matter' and I have called a 'porosity theory'. Although Shapiro gives credit for this theory to Robert Boyle, a similar theory underpins Willis's treatment of structural composites, especially in his tract on fermentation included in *Diatribæ medico-philosophicæ* [1659].

[23] Willis, *The Anatomy of the Brain* [1664 tr. 1681], p. 125.

In the second-century the physician, Galen, named this structure *chiasma* because of its likeness to the letter X, the Greek term for which is *chi*. Here is his description of the structure of the optic nerves, which proceeds from the brain to the eye (the opposite direction of Newton's description):

> ...the two offshoots [the optic nerves] proceed from this area [of the thalamus[24]] and come to [the anterior or lateral ventricles in] the middle region [of the brain], where the one offshoot connects with and joins the other. Then from that place each one of the two separates off and leaves the other, and they both proceed obliquely to the region of the eyes; and so their form is that of the Greek letter "chi", which is written "X". Neither one of the two offshoots passes to the side opposite that of its origin, but rather the nerve which originates on the right side of the brain goes to the right eye, and the nerve which originates from the left side of the brain proceeds to the left eye.[25]

There are two points to note about Newton and Galen's descriptions.

First, although both men provided a clear description of the X-shaped structure of the optic nerves, yet there is a notable discrepancy between their accounts. For in Newton's description from the back of the eye to the brain, we are told that the optic nerves 'meet' *before* they come into the brain, whereas in Galen's description from the brain to the back of the eye, we are told that the optic nerves 'join' when they come to the anterior or lateral ventricles in 'the middle region' of the brain. It should be noted, therefore, that Newton's description, though insufficiently detailed, is more or less correct. For today we know that the optic nerves on opposite sides behind the globe of the eye are connected together at the region of their union or commissure at the base of the brain; and from the back of that commissure, they may be traced into the brain under the name of optic tracts.

Second, since Galen regarded the optic nerves as visibly hollow,[26] he made no mention of fibres, when in *De usu partium* he wrote that

> ...certainly she [Nature] did not interchange them [the optic nerves], bringing the one from the right side to the left eye and the one from the left side to the right eye; yet the shape of these nerves does greatly resemble the letter Chi (X), and some one whose dissection was not accurate might think, perhaps, that they are interchanged and do cross one another. *But this is not the case*; for after they have met inside the cranium and their channels[27] have

[24] Galen's term 'thalamus', which occurs in only one passage of the Greek corpus of his writings, denoted the recess (now the inferior cornu) in the lateral ventricles of the brain which, in his anatomy, was the source of the optic nerves; see Smith, 'Galen's Account of the Cranial Nerves', pp. 85–6 and 96 n.34. Note, therefore, that his anatomy of the thalamus differs from that of the two seventeenth-century physicians treated *infra* Pt.III.3.1 and III.3.2.

[25] Ibid., pp. 86–7 (additions in square brackets mine). According to Koelbing, 'Ocular Physiology', pp. 219–20, Galen's knowledge of the anatomy of the eye, including what now is called the optic tract, was 'very accurate'; and this judgment was reaffirmed by Spiegel, *Galen on Sense Perception*. For some continental writers (but not English ones) who noticed or described the X-shaped structure after Galen, see Crombie, *Science, Optics, and Music*. But the term '*chiasma*' (or its English equivalent, 'chiasm') did not become a term of art in English anatomy until the nineteenth century (see OED).

[26] See Smith, 'Galen's Account of the Cranial Nerves', pp. 85–7, p. 86.

[27] According to Galen, Herophilus of Alexandria had referred to the optic nerves as channels' (*poroi*); see May (tr.), *Galen on the Usefulness of the Parts of the Body*, Bk. X, Ch. 12, p. 491.

> united, they at once draw apart again, showing clearly that there was no other reason for them to approach one another save the joining of their channels.[28]

So also Kepler, who in 1604 wrote only of a 'union of images (*species*)' at the 'junction' (*conjunctio*) of the optic nerves, the latter of which he regarded as 'hollow' conduits for the spirits. He also pointed out that these nerves were not optically in a direct line, for

> ...if any nerve went to its seat in the brain freely in a straight line, with two eyes we would think we saw two things instead of one. Either this junction takes place so that when one eye is closed this hidden seat in the brain should not cease from its function of judging. Or perhaps the actual doubling of the seats is not only on account of the eyes, but is for the purpose of correct judgement of distances, as with a pair of eyes. Therefore, in order that visible things may be judged correctly and a distinction made between what is seen with one and with two eyes, this junction of the passages must take place. [*Translation by A. C. Crombie*][29]

Unfortunately, the position of Newton, who does mention the fibres, is not altogether clear. For in the remaining part of his convoluted sentence, remarked on above, he indicates that before the optic nerves enter the brain, 'Species of Objects seen with both Eyes' are 'united' in 'the same place' where the optic nerves 'meet'. But after 'union', the nerves separate as they go into the brain. And in the brain, one nerve goes to the left and the other, to the right side of the *sensorium*, where the fibres 'make but one entire Species or Picture', half of which on the right side of the *sensorium* comes from the right side of both eyes '*through* the right side of both optick Nerves to the place where the Nerves meet' and half of which on the left side of the *sensorium* comes in the same manner from 'the left side of both Eyes'.

Some commentators claim that this passage provides evidence for Newton's awareness that the fibres cross.[30] But this claim is problematic for several reasons. In the first place, Newton uses the terms 'meet', 'unite' and 'separate' rather than the terms 'cross' or 'decussate'. In the second place, his description of the coats of the optic nerves as 'cylindrical' suggests that their fibres are not, like a pack rope, twisted from many fibres but rather that the fibres are internal to the surfaces of the nerves. In the third place, only when microscopes had become sufficiently powerful was it possible in 1781 to identify *microscopically* the fibre as the neuron of the axon, so that, according to Edwin Clarke, naked-eye observation 'presumably' resulted in identifying *macroscopically* a bundle of axons as the fibre.[31]

---

[28] Ibid. (italics mine).

[29] See Crombie, *Science, Optics and Music*, pp. 298–9.

[30] See, e.g., Boring, *A History of Experimental* Psychology, pp. 75, 675, and Boring, Sensation *and Perception,* pp. 227 and 257. Note that this author assumed that Galen, as well as Kepler explained singleness of binocular vision in terms of the 'semi-decussation' of the fibers at the junction ('chiasma') of the optic nerves.

[31] See Clarke, 'The Doctrine of the Hollow Nerve', pp. 124, 137–9. It is possible that this was the result of the naked-eye observations of Andreas Vesalius, who is said to have identified macroscopically the existence of peripheral nerve fibre; see Clarke and Jacyna, *Nineteenth-Century Origins of Neuroscientific Concepts*, p. 388 n.10.

In the fourth place, and of importance, seventeenth-century anatomists, including Thomas Willis, commonly held that fibres (threads, filaments) are everywhere in the human body. And Willis also held that fibres are 'every where alike', thereby denying their specificity, a conception that emerged much later in the nineteenth century. Consequently, he explained all 'sensible' difference as due to the manner in which fibres are 'irritated', so that 'the Animal Spirits implanted in them' are stimulated to enter into 'a peculiar way of Gyration or turning round, or of undulation or waving'.[32] Fifth and finally, although there seems to have been some discussion in the seventeenth century concerning the decussation of fibres in the optic nerves,[33] Newton merely claims that after 'union' the two optic nerves together with their fibres terminate on either side of the *sensorium*. Accordingly, until further investigation of this issue is undertaken (and this is beyond the scope of my inquiry), I infer that Newton was unaware that fibres decussate.

To return, then, to the convoluted passage from Text VII (Qu. 15), it is clear that Newton here raises the problem of singleness of vision, a phenomenon now called 'binocular fusion', namely, how we see one object with both eyes despite the retina of each eye having its own image[a] of the object. After Kepler's optical analysis of the formation of images on the retinal coat, the optical processes of binocular fusion were explained by the hypothesis that an object is seen single because its images fall on corresponding points of the two retinas, the implicit assumption being that images projected on each retina are exactly similar.[34] This hypothesis was adopted by Descartes[35]; and it also was the dominant hypothesis until the nineteenth century. Indeed, about the middle of that century, William Whewell, in a chapter on aphorisms concerning ideas, claimed that:

> *The* Second Paradox of Vision *is, that we see objects single, though there are two images on the retinas, one in each eye. The explanation is, that it is a Law of Vision that we see (small or distant) objects single, when their images fall on corresponding points of the retina.*[36]

But Whewell notwithstanding, in 1838 Charles Wheatstone, using experiments with his new-invented optical instrument, the Stereoscope, was already bringing into disrepute the hypothesis of corresponding points as an accurate explanation of the phenomenon of binocular fusion.[37]

---

[32] See Willis, *Two Discourses*, p. 61. See further, also for the source of fibre irritation, *infra* Pt.III.3.2, p. 79 and n.25.

[33] See *infra* Pt.III.3.1, pp. 64–5.

[34] According to Boring, *Sensation and Perception*, pp. 226, 257, this hypothesis was first mooted in 1613.

[35] See Crombie, *Science, Optics and Music*, p. 235 n.108.

[36] See Whewell, *Novumn Organon Renovatum*, p. 17.

[37] See Wheatstone, 'Contributions to the Physiology of Vision', p. 389. Note that Wheatstone's experiments with the Stereoscope also drew attention to a new phenomenon now called 'steropsis', in which there is a separation between objects as seen by the left and by the right eye (retinal disparity). Investigations into this phenomenon gradually led to the understanding that these differences between the left and the right eye provide information that the brain can use to calculate depth in a visual scene. Note also that Wheatstone's Stereoscope was the outcome of his explora-

It would not be surprising, therefore, to find Newton himself holding the hypothesis and its underlying assumption. Indeed, about 1666 he certainly accepted it, writing that 'the pictures of any object ought to bee made upon the corresponding points of ye two Retinas'.[38] And although Text III implies that acceptance, yet in Text VII (Qu. 15) he is concerned with a visual problem, not an optical process, the former of which he explains physiologically as arising from the anatomical structure of the optic nerves and the continuity of their fibres, both of which are conduits, not agents of visual processes.

In 1664 the problem of singleness of vision had been addressed by Thomas Willis, who traced the anatomical pathway of the optic nerves from the brain to the back of the eye, like Galen's procedure but the reverse of Newton's. He then appended a brief statement concerning the 'offices' and use of the X-shaped structure of the optic nerves as follows:

> The growing together of these Nerves, and their being again separated, seems to be ordained for this end, that *the visible species*, received from either Eye, might appear still the same, and not double; for this conjunction of the twofold Organ [the optic nerves] frames the double image into one, which once united [at the conjunction], ... [the image[b] is afterwards] carried to either side of the common Sensory ... [where] it is on both sides alike, [and] appears still the same [i.e., one, not double]. [*Translated by S. Pordage*, italics mine].[39]

The above passage, as rendered by the translator, consists of one long sentence broken by a semicolon, after which there are a number of difficulties in the translation that cloud its meaning. If my editorial efforts at clarification are correct, then the first part of Newton's explanation of singleness of vision is somewhat similar to that of Willis, with the exception that, after union, the image[b] to be transmitted is halved and becomes whole only when the entities contained in the fibres stimulate the two sides of the *sensorium*. But the second part of Newton's explanation may have been motivated by Willis's requirement that the fibres in the optic nerves, 'by a direct beam', carry the 'visible Species, sufficiently refracted in the Eye' to 'either side of the common Sensory'.[40] This term, the Latin of which is *sensorium commune*, was the subject of increasing debate from the sixteenth century.[41] But, as will become evident later on, Willis's dissections of the brain led him to reject all previous theories and to replace them with a new theory of his own.[42]

---

tion of the analogy between sound and light, which led him to a study of vision and an understanding of the principles of binocular vision; see Bowers, *Sir Charles Wheatstone*, p. 3, *et passim*. For a brief historical discussion of stereoscopy, see Boring, *Sensation and Perception*, pp. 282–88.

[38] See Newton, 'Of Colours', p. 485.

[39] See Willis, *The Anatomy of the Brain* [1664 tr. 1681], p. 103. Newton would have had access to Barrow's copy of the first edition of this book; see Feingold, 'Isaac Barrow's Library', (No. 1048), p. 369. For evidence that he read it, see *infra* Pt.III.3.2, p. 80 and n.106.

[40] Ibid., p. 139.

[41] For some of the anatomists involved in the debate prior to Descartes, see Hall in Descartes, *Treatise on Man*, p. 40 n.72. As pointed out *supra* Introduction, pp. xvii, xix, he did not use the term *sensorium commune* in this work or in his *Discours de la méthod.*

[42] For details, see *infra* Pt.III.3.2.

After Willis, a second English physician also attempted to explain why with two eyes vision is not double; and it is this physician who is implied in Text VII (Qu. 15). For in the concluding sentence of that text, Newton alludes to an opinion concerning warm- and cold-blooded animals, the former of which have the X-shaped structure of the optic nerves, whereas the latter do not, after which Newton adds: 'if I am rightly informed'. The clue to his informant is to be found in the instances he gives of cold-blooded animals, namely, 'Fishes' and 'the Chameleon', for in a tract published in March 1682, William Briggs noted that

> In many *Fishes* ... the two [optic] Nerves are joined only *per simplicem contactum*, and in the *chamæleon* not at all (as is said) though I never had the opportunity of making any inspection into that Animal, and therefore can say nothing of it (no more than I did *formerly*).[43]

And in September of the same year, Newton wrote to Briggs concerning the problem of singleness of vision.[44]

Turning now from the visual sensory to the motor system, only two texts—Texts II and X—include a consideration of 'animal motion', in connection with which Newton avoids using the term '*sensorium*'. Instead, he merely observes that the anatomical pathway for the motor system is from the brain through the nerves to the muscles. But this consideration is not the only feature that differentiates the two texts from those on the sensory system, for they also include three other distinctive features. First, Newton uses the term 'capillamenta', not fibres. Second, in Text II he posits unobservable ætherial substances in the pores, as well as outside the surfaces or coats of the anatomical substrates or containers, whereas in Text X he posits an ætherial medium in the pores of the containers. Third, and perhaps the most important, he conceives animal motion as a voluntary action determined by a soul and, hence, offers an alternative to the mechanistic physiology of Descartes (who is not named).[45]

It is worth pointing out that in a manuscript written during the same period as Text II, Newton included a brief consideration of the problem of animal motion in relation to the motor act. This consideration occurs chiefly in a critique of Descartes's conception of the nature of space as the geometrically extended impenetrable body dispersed everywhere. To counter that conception, Newton attempted to deduce the nature of compound matter from 'our faculty of moving our bodies' and thus to reduce his conception of 'body' to that faculty. In summarising his own 'deduced' conception, he wrote as follows:

---

[43] Briggs, 'A New Theory', p. 170. For what he said '*formerly*' concerning the formation of the eye in different animals (in which there is no mention of the X-shaped structure), see Briggs, *Opththalmo-graphia* [1676], Ch. VII, pp. 66–72. Concerning the 'union' of the optic nerves in fish, see also Willis, *The Anatomy of the Brain* [tr. 1681], p. 77. According to Mendelsohn, *Heat and Life*, p. 40, not until the seventeenth century did comparative anatomical studies between humans and other animals become essential.

[44] See *infra* Pt.III.3.1.

[45] For the premiss of that physiology, see *infra* Pt.III.3.2, p. 81.

> ...the power of creating minds is not delineated in any faculty of created mind ... [so that] in moving bodies we create nothing, nor can we create anything, but we only simulate the power of creation. For we cannot make any space impervious to bodies, but we only move bodies; and at that not any we choose, but only our own bodies, to which we are united not by our own will but by the divine constitution of things; nor can we move bodies in any way but only in accord with those laws which God has imposed on us.[46]

How, then, do we move our own bodies? A partial answer is to be found in Text II,[47] where Newton treats animal motion as a problem of muscle action. However, before he considers this problem, he presents, and then dismisses three suppositions concerning it. In the extracts that constitute Text II, I have not included his presentation, the perspicuity of which makes it difficult to identify with any certainty the authors of each supposition.[48] Nevertheless, all three suppositions focus on a motive 'power' of the soul to contract the muscle 'at will', and this suggests an indebtedness to Galen's natural philosophy.

According to that philosophy, the body is wholly in the service of the soul, and the spirits (*pneumata*) are instruments or material causes which the soul avails itself in the exercise of its functions. Indeed, all three suppositions seem to represent variants of a theory found in Galen's treatise on muscle physiology, in which he taught, among other things, that true muscles are exclusively organs of voluntary action and that muscle contraction is brought about by animal spirits.[49] These spirits, which are stored in the brain's ventricles, undergo a threefold refinement before passing from the fourth ventricle through the nervous system to the muscles,[50] so that animal motion is produced when a muscle as a whole increases transversely and at the same time shortens longitudinally.

In 1522 a Latin translation of the treatise was published in London by the physician, Thomas Linacre, a pupil of the translator. And this publication in turn led to numerous modifications of Galen's theory, examples of which include the three variants noted above, as well as Newton's own variant.[51] Before describing his variant, it is important to point out that the extracts on animal motion presented in Text II do not make clear that Newton's solution to the problem of muscle action presupposes a porosity theory of the nature of compound matter; and this needs to be remembered when encountering words such as 'æther', 'juice', 'spirit' and the like. What Text II does make clear is that to solve the problem of muscle action, he relies on suppositions that are mechanical, as well as proto-chemical.

---

[46] Newton, 'De gravitatione', p. 141.

[47] See also *infra* Pt.IV.4.2, pp. 107–8.

[48] For the omitted text that contains the three suppositions, see Cohen and Schofield (eds.), *Newton's Papers*, pp. 182–3; see also *supra* Pt.I.1.2, Text II, p. 6 n.18.

[49] For Galen's theory, see Bastholm, *The History of Muscle Physiology*, pp. 74–96.

[50] See Woollam, 'Concepts of the Brain and its Function in Classical Antiquity', pp. 17–9.

[51] For some seventeenth-century variants of the theory, see ibid. Bastholm (n.49 above); see also Needham, *Machina Carnis* and Fulton, *The History of the Physiology of Muscle*. Although important contributions to muscle physiology were made in the seventeenth century, yet physiological thought was dominated by the belief in animal spirits until the nineteenth century.

The mechanical suppositions may be instanced from his analogy of nature between the properties of a (hypothetical) æther and the air. This analogy is indebted to Robert Boyle's experiments concerning the compression (density/weight) and dilation (spring/pressure) of the air.[52] But in applying this model to the problem of muscle action, Newton conceives the elasticity of the æther to be stronger than that of Boyle's air. He then supposes that the pressure of the æther included *within* the coat of the muscle is in 'counterpoise' with that of the æther *outside* the coat of the muscle. He further supposes that by means of a change of 'consistency' of the included æther, the muscle increases in volume, that is to say, on contraction the muscle 'swells'.

To account for how this change in the included æther brings about contraction, Newton introduces at least two proto-chemical suppositions. According to him, the material cause of change is an ætherial animal spirit sent from the brain into the muscle through the nerves. This 'third' ætherial entity is a 'mediator' between the æther outside the coat of the muscle and the 'juices' within it. For by its action, the ætherial animal spirit makes those juices 'more sociable' to the exterior æther, thereby allowing the latter to 'pervade' the muscle so that the two entities (i.e., their particles) mix more freely. In short, the 'mediator' acts like a solvent,[53] namely, a substance that has a chemical power to dissolve, and hence open up the parts and pores of the matter on which it works.

Newton then states that although 'what is said of muscular motion, may be applied to the heart', there is 'this difference'. And in describing this difference he introduces a second proto-chemical supposition. For he claims that the heart contains a 'ferment', the unspecified action of which results in a 'fermentation' of the nourishing 'juices' that replenish the flesh of the heart. And this fermentation 'raises' an ætherial vital spirit, which is sent 'by starts' from the heart into the brain through 'some convenient ductus'. By inference, therefore, we must conclude that the ætherial vital spirit is converted into the previously-mentioned ætherial animal spirit by some undisclosed chemical process in the brain substance. For all muscles, the heart included, receive the ætherial animal spirit 'by impression' from the brain; and the internal cause of that impression is a soul which has a power to determine the 'ætherial animal spirit or wind into this or that nerve' and thus to cause 'all the motions we see in animals' (see also the related Text X, Qu. 24).

Despite some similarities, Newton's conceptions cannot be described as Cartesian, in part because he has recourse to two kinds of spirits, animal and vital, whereas Descartes reduced spirits to animal only.[54] Therefore, in order to increase

[52] For an account of Boyle's experiments on the compression and dilation of air, see Webster, 'The Discovery of Boyle's Law', pp. 484–90.

[53] Note that Willis devoted two chapters (IX and X) to solvents in his tract on fermentation included in *Diatribæ duæ medico-philosophicæ* [1659]; see also *supra* Pt.I.1.2, Text II, p. 7 n.24.

[54] See Descartes, *Discourse on Method*, pp. 37–48, for the heart, the blood and the production of animal spirits, which, 'like a very subtle wind, or rather like a very pure and lively flame', arise 'continuously and in great abundance from the heart to the brain' and go from there 'through the nerves into the muscles' to impart 'movement to all the members'. For details of Descartes's physiology, see, e.g., Davis, *Circulation Physiology and Medical Chemistry*, pp. 40–6, 101–108, 112, *et passim*.

our understanding of his text, we need to discover the meaning of such terms as 'capillamenta', 'juices', 'ferment' and 'fermentation'. Let us begin with 'capillamenta', which, as previously noted, is unique to Texts II and X.[55] In 1664 this term was coined by Willis to denote fibres in the nerves,[56] but he also used the same term for the 'little hairs of Arteries and Veins, together with their proper Fibrils'.[57] It is probable, therefore, that he extended the term 'capillamenta' to fibres on the model of the capillary vessels—the extremely minute blood vessels, in which the arterial circulation ends and the venous begins.[58] Indeed, evidence for this probability appears in his description of the '*Fibrarum genus & natalita*', where he treated both embryonic nerve fibres and capillary vessels as rudiments. At conception, these rudiments, which are already in the embryonic brain and heart, develop slowly into the nervous and circulatory systems if properly 'watered' (*irrigari*) by certain 'juices'.

Willis identified two juices that are distributed by the 'primigenious parts'—the heart and brain with the arteries and nerves 'hanging to them'. One juice, a thick, oily 'nutritious juice' (*succus nutritis*), is a serous humour deposited throughout the arteries from the bloody mass', whereas the other, a spirituous 'nerve juice' (*succus nervosus*), is a nervous 'liquor' that diffuses with itself 'very active Particles' perpetually flowing, though but in small quantity, from the brain through the nerves.[59] For animal life, sensation and motion, the nerve juice 'couples with' the oily nutritious juice, the latter of which is then 'actuated' by the former. And, like 'a certain Ferment', this '*Copula*' or mixture results in a chemical reaction in which the blood boils and releases a vital spirit in the heart. Accordingly, the *Copula* is the material cause of the boiling motion (ebullition) that Willis called 'fermentation'.

The steps in the chemical process described above may be illustrated briefly with reference to Willis's conception of the heart,[60] the exterior of which he conceived as a blast bellows and the interior, as an open hearth or forge ('fireplace') in which a chemical reaction takes place. This reaction 'enkindles' the blood,[61] so that it regains

---

[55] For this term, see also *supra* Pt.I.1.2, Text X and comment.

[56] See Willis, *Cerebri anatome* [1664], Ch. XIX (on the nervous system in general), p. 246. In English the term appears as 'capillament', see Willis, *The Anatomy of the Brain* [1664 tr. 1681], p. 128. See also OED obs. (1681–1751), where the first citation is to the 'Table of all the hard words derived from the *Greek* and *Latin*', appended by the translator, Pordage, to a collection of some of Willis's writings published in 1681. See also Briggs, *Ophthalamo-graphia* [1676], in which the description of one of his figures (Fig. 5) refers to '*Nervi Optici capillamenta*'.

[57] Willis, *The Anatomy of the Brain* [tr. 1681], p. 89.

[58] See Willis, *Cerebri anatome* [1664], pp. 246–8, and Willis, *The Anatomy of the Brain* [1664 tr. 1681], pp. 128–9.

[59] Willis, *The Anatomy of the Brain* [1664 tr. 1681], pp. 129; see also Ch. XX ('Of the Nervous Liquor, and whether that or the bloody Humor be Nutritious'), pp. 131–6. Note that the two 'juices' represent Willis's version of the two-semen theory of Hippocratic tradition.

[60] For Aristotle's treatment of the heart in 'Parva naturalia', see Hett (tr.) *Aristotle on the Soul*, XX–XXI, pp. 475–81, especially p. 479, where he compared the shape of the heart to 'both the bellows in a forge'.

[61] For 'inkindling' ('infiring', 'accension'), see Willis, *The Anatomy of the Brain* [1664 tr. 1681], pp. 95, 126, 129, 131, 136, 152, 154, 162.

the heat for its journey by the arteries that it lost when returning to the heart by the veins. During enkindling, the blood 'boils up'[62] and releases heat as a vapour called 'vital spirits', which, in turn, 'blows up' (inflates) the heart. Accordingly, he equated the heart's contraction with filling (inflating) and its relaxation with compressing, when, like a blast bellows, the heart drives out the stream of blood.[63] By these two mechanical motions, the pulsations of the heart's two halves maintain the blood's flow at constant pressure as well as constant temperature.

But Willis also conceived the blood vessels as distillatory organs,[64] in which the circulating blood undergoes a sequence of chemical processes during which the 'purer and more active particles' of the blood are separated out and further subtilised until, like 'a Chymical Elixir',[65] they reach their final 'perfection' in the cortex of the brain. Accordingly, the perfected particles called 'animal spirits' are the brain's 'effluvia',[66] that is, a flowing out of real material particles too subtle to be perceived. And thus, as Newton points out in Text II, they are not the same as 'the liquor, vapour, or gas of spirit of wine'.[67]

As for the motion of 'the Muscles of the whole Body', Willis stated that they have 'a certain Analogy with the motion of the Heart'.[68] But he rejected what William Harvey had previously demonstrated, namely, that the heart's 'pulsific force' (its contractility) is intrinsic to muscle fibre.[69] Against this, Willis maintained that the motive power is *inside* the pores of fibres in the muscles, so that on contraction, the heart and other muscles inflate, that is, increase in volume. In 1664 Willis several times compared the chemical reaction that leads to muscle inflation with the elastic force of gunpowder, which, when 'rarified and inkindled', produces

[62] For 'boiling up' ('ebullition', 'deflagration'), see ibid., pp. 84, 90, 96, 104, 124, 131, 155.

[63] Ibid., p. 135.

[64] For the brain as an alembic, see ibid., pp. 87–8, *et passim.*

[65] Ibid., p. 88.

[66] Ibid., p. 77.

[67] Cf. Willis, *Two Discourses* [1672 tr. 1683], pp. 23–4: 'The comparing of these [animal spirits], with the Spirits of Wine ... and such like, does not quadrate or agree [with their essence]. For besides, that ... Chymical Liquors, neither represent the Images of their Objects, nor are indued with any Elastic Virtue as the Animal Spirits; those also are less Subtle than these'. For the motive power called 'Elastic Virtue', see further below.

[68] Ibid., p. 135.

[69] See the first tract in Harvey, *The Anatomical Exercises*, pp. 46, 69–70, in which he clearly recognised that contractility is a relation between stretch ('constriction being oblong') and tension. Note that Frank, Jr., *Harvey and the Oxford Physiologists*, did not grasp that the Oxonians' interest in muscle action began with concepts from Galenic tradition, for this tradition better suited their proto-chemical interests.

an 'explosion'.[70] Then, in his 1667 book on 'Convulsive *Pathology*',[71] he devoted the first chapter to describing the action of healthy muscles, in which he concluded that

> ...the motive virtue [i.e., power], both of the Heart, and the rest of the Muscles, becomes more strong and Elastick, above any mechanick Organ ... [when] the animal spirits, acting every where in the musculous stock, get to themselves an explosive *Copula*.[72]

And, in the paragraph immediately following, Willis claimed that the word '*Explosion*' had yet to be used either in natural 'Philosophy or 'Medicine', thereby implying he was the first to do so.

Since Newton owned a 1668 edition of Willis's 1667 book,[73] it is plausible to assume that, as a Fellow of the Royal Society, he purchased his copy after reading a review of the original edition in the Society's *Philosophical Transactions*.[74] For according to the review, Willis

> ...teacheth that there are indeed *Animal* Spirits, that they constitute the *Being* of the *Corporeal* Soul, and are also the next and immediate Instruments [i.e., material causes] of all Animal motions, producing them by a kind of *Explosion* or Shooting, upon which *Elastick* or *Explosive* Power, he establishes his whole Doctrine of *Convulsions*.[75]

On this succinct representation of Willis's teaching, we might conjecture that it constitutes one of the three suppositions concerning muscle action mentioned earlier in this section, the third of which Newton puts as follows: 'the soul has a power to inspire any muscle with this [ætherial animal] spirit, by impelling it thither through the nerves'.[76] But he rejected this supposition, because it required too strong a force and, consequently,

> ...a forcible intending[77] the spring of the æther in the muscles, by pressure exerted from the parts of the brain: and it is hard to conceive, how so great a force can be exercised amidst so tender matter as the brain is. And besides, why does not this ætherial [animal] spirit, being subtil enough, and urged with so great a force, go away through the dura mater[78] and skins [i.e., coats] of the muscle; or at least so much of the other æther of the muscle go out to make way for this, which is crouded in?[79]

---

[70] See Willis, *The Anatomy of the Brain* [1664 tr. 1681], pp. 129, 135, 152–3. In an early tract, he devoted a chapter on nature's solvents (fire and water) that included a description of gunpowder and its effects; see Willis, *A Medical-Philosophical Discourse of Fermentation* [1659 tr. 1681], pp. 41–2.

[71] I.e., Willis, *Pathologiæ cerebri, et nervosi generis specimen.*

[72] Willis, *An Essay of the Pathology of the Brain* [1667 tr. 1684], p. 2.

[73] See *supra* Pt.I.1.2, comment to Text II.

[74] *Philosophical Transactions of the Royal Society* (1667), 2 (No. 31), pp. 600–2.

[75] Ibid., p. 601.

[76] See *supra* Pt.I.1.2, Text II, p. 6 n.18.

[77] See OED *vbl. sb.*, rare: The action of stretching.

[78] I.e., the fibrous membrane forming the outermost covering of the brain.

[79] See Cohen and Schofield (eds.), *Isaac Newton's Papers*, p. 183.

As will appear in Part III, sometime before December 1675 (the date of the first part of Text II), Newton was in contact with William Briggs, who, in a treatise published in the following year, offered an opinion that a body in a healthy condition would be adversely affected by a violent explosion of spirit ('violenta spirituum explosio'), as well as by a fermentation arising from the mixture of the liquor in the nerves and blood ('vel fermentatio illa à nervei & sanguinei liquoris mixtione orta').[80] In this opinion he criticised the two previously examined concepts that Willis had attributed to chemical reactions. Instead, he offered his own mechanical conjecture as better suited to explain the manner in which the ocular muscles contract. His conjecture is based on two suppositions. First, the whole body of the nerve, including its fibres, is watered (*irrigari*) by a continuous influx of animal spirit, which, in equal amount, is supplied by the nerves to the muscles. Second, by a motive power of the principle of life (*animæ imperio*), a slight quantity of spiritous juice (*succi spirituosi*) is propelled into one rather than another fibre to cause contraction (i.e., inflation) of the supplied muscle, thereby overcoming its antagonist in the same way that a slight weight, added on one side of a balance in equilibrium, imparts a movement to that side.[81]

In 1670, having been convinced by his former assistant, Richard Lower, that the heart had no innate ferment,[82] Willis modified his conception of a chemical reaction in the heart. Instead, he concluded that the body's heat must come from a chemical reaction in the blood, a reaction he called an 'accension'.[83] Briggs's criticism, therefore, indicates that in 1676 he was aware of Willis's modified position, whereas in Text II Newton's reference to 'the ferment in the heart' provides evidence that in 1675 he was unaware of it.[84] Moreover, Briggs's use of the terms '*irrigari* and *succi spirituosi* suggest that he retained a number of Willis's other concepts,[85] whereas his mention of '*animae imperio*' in connection with muscle action is reminiscent of

---

[80] Briggs, *Ophthalamo-graphia* [1676], p. 48.

[81] Ibid., pp. 46–8. Note that Briggs's mechanical analogue is a balance, which is analogous to a lever, the analogue Willis used in his 1670 tract, 'De motu musculari', which, with two other tracts, was included in Willis, *Affectionum que dicuntur ... pathologia spasmodica vindicata*, in which the myology of Nicolaus Steno is mentioned with respect, even though it contained criticisms of some of Willis's ideas.

[82] In 1669 Lower concluded from experiments that the motive power is due not to an explosion but to an orderly inflow of spirits into the muscles through the nerves; for full details, see Lower, *Tractatus de Corde*, pp. 62–86, *et passim*. For a review of his tract, see *Philosophical Transactions of the Royal Society* (1669), 4: 909–12. According to Bastholm, *The History of Muscle Physiology*, p. 148 n.2, Lower made use of Steno's illustrations of the heart as a muscle.

[83] OED obs. or arch.: The action of kindling or the state of being kindled; ignition; inflamation; heat. For his tract on the accension of the blood, see Willis, *Affectionum que dicuntur ... pathologia spasmodica vindicata*,

[84] By *c*.1687, however, he was aware of it. See *supra* Pt.I.1.2, comment to Text X, p. 24, where he writes that 'the inrush of animal spirits' does not swell the muscles, since they were 'already full before'; rather, contraction is due to the micro-mechanical interchange of particles (i.e., the motion of fermentation that generates heat), and 'when this ceases', the muscles 'grow flaccid'.

[85] E.g., Willis's *succus nervosi* and *succus nitritis* mentioned earlier in this section; for their equivalent, see Briggs, *Ophthalamo-graphia* [1676], pp. 39 and 43.

Newton's treatment in Text II of the motive power that produces the effect of contraction.

According to that text, 'the soul', from its undisclosed place in the anatomical space in the brain, causes 'all the motions we see in animals', that is to say, as the efficient cause it determines 'at will' the ætherial animal spirits into the requisite nerves that innervate a muscle or muscles. Because these spirits, which he also refers to metaphorically as 'wind',[86] are perfected in the brain substance from ætherial vital spirits generated in the heart, we may conclude that the whole distributed body of immovable ætherial spirits, vital and animal, is Newton's *corporeal* principle, that is to say, the material cause of animal life, sense and motion. The soul of Text II is a different principle,[87] because, on the one hand, it is immovable and, hence, not distributed; and, on the other hand, it is the efficient, not material cause of animal motion that is voluntary. By implication, therefore, this soul is Newton's *incorporeal* principle of intellectual life. In what follows, I shall designate this latter principle as *animus* and the former principle as *anima*, even when I refer only to that part of *anima* concerned with animal spirits, rather than vital spirits.

2.2 Summary and Comment

From the brief comments to Newton's texts in Pt. I.1.2 and from the analysis of some of these texts in the previous section, it now is possible to provide a summary of Newton's conception of the sensory-motor system and its relation to the human *sensorium*. But before doing so, it is important to recall that he conceived the nervous system on a container/contained model. According to his version, the containers, which have different sizes and shapes, are dense bodies with pores, the latter of which (the pores) contain unobservable and, hence, hypothetical corporeal substances of different sizes and chemical powers (e.g., dispositions to congregate and separate). Some of these particulate substances (e.g., ætherial animal spirits) fill the pores of the containers in the nervous system, whereas others (e.g., rays of light[88]) are introduced into the pores of some of those containers. In short, the containers are the anatomical substrates for sensory-motor functions, and the contained substances perform those functions by means of a more or less vigorous micro-mechanical interaction, like the motion of bubbles when water is boiled.

In his texts relating to the human *sensorium*, Newton uses terms such as vibration, trembling motion and agitation to denote this micro-mechanical interaction of particles in the pores of dense bodies. But it could be inferred that all these terms are

---

[86] See *supra* Pt.I.1.2, Text II. For Descartes's use of this metaphor, see p. 40 n.54 above; for Willis's extended metaphor, see *infra* Pt.III.3.2.

[87] The same is true of the soul of Text IX; see *supra* Pt.I.1.2, comment to Text IX.

[88] See *supra* Pt.I.1.2, where according to Texts II light rays enter the eye and strike the coat of the retina and according to Text VI 'the several sorts of rays ... by mixing their actions in the Sensorium beget a sensation' of colour.

synonyms for the motion he calls 'fermentation'. Indeed, elsewhere he identifies this motion as a vital process of animal life, because the 'active Principles' that produce fermentation keep 'the Heart and Blood of Animals ... in perpetual Motion and Heat'.[89] In the sensory system the micro-mechanical motion of fermentation is excited externally by particles that strike the receptors of the external senses, whereas in the motor system the micro-mechanical motion is excited internally by an unexplained entity called 'the soul' (*animus*).

As for the unobservable substances that are in micro-mechanical motion, Newton offered a number of different assumptions about their nature, for example, conceiving them as ætherial spirits,[90] as absolutely hard particles (atoms or *corpuscula*)[91]; or as electrical spirits.[92] I have intentionally avoided entering into detail about his changing assumptions about these hypothetical media, because, on the one hand, to do so would take me beyond the scope of this inquiry, and, on the other hand, the substances themselves, as well as Newton's methodological treatment of them have been examined in detail by Alan Shapiro.[93]

Bearing in mind, then, the general points made in the previous paragraphs, the following summary begins with the visual sensory system, which is Newton's paradigm for the sensory system in general. Consequently, he focuses on visual function.

First (Texts II, IV–VI, VII Qu. 12, X), the human *sensorium* is one of the containers in the nervous system, the location of which is some unspecified place in the anatomical space of the brain.

Second (Texts III, VIII, XI), there is an implicit analogy of nature between the retina in the interior space of the eye and the *sensorium* in the interior space of the brain, since each contain in their pores an unobservable corporeal substance, the property of which is sensitivity.

Third (Texts I, III, VII Qu. 15, XI, XII), both sensitive substances are adapted to certain kinds of stimuli, that in the retina for pictures (images[a]) and that in the *sensorium* for sensible species (images[b]); accordingly, the retina is the peripheral end-organ and the *sensorium*, the central end-organ for vision.

---

[89] See Newton, *Opticks (C-R)*, Query 31, p. 399.

[90] See *supra* Pt.I.1.2, Text II and comment, and Text X and comment.

[91] See, e.g., *supra* Pt.I.1.2, Text IV and comment.

[92] See Newton, *Principia mathematica (C-M)*, Bk.III, General Scholium, p. 547, and Newton *The Principia*, Bk.III, General Scholium, pp. 589–90. For his reliance on the last medium to elucidate sensory-motor function, see Newton, 'De motu et sensatione Animalium" and its companion text, 'De vita & morte vegetabili', which pre-date by two or three years the composition of the General Scholium that concludes the second 1713 edition of *Principia mathematica*. Note that at the end of the comment on this manuscript, the transcribers add a quote from f. 241 of a manuscript that is not part of the texts transcribed. The quote is as follows: 'This [electrical] spirit therefore may be the medium of sense & animal motion & by consequence of uniting the thinking soul & unthinking body'.

[93] See all of Part I in Shapiro, *Fits, Passions, and Paroxysms*; see also Shapiro, 'Newton's Optical Theories and Vibrating Media'.

Fourth (Texts II, III, VII Ques. 12 and 15, VIII, XI), vision requires an optical, as well as a visual process: the first process begins when external stimuli called rays of light strike the coat of the retina, thereby exciting in its pores a light-sensitive substance; the second process begins when the light-sensitive substance mixes with, and excites an unobservable corporeal substance in the pores of the fibres that exit behind the retina.

Fifth (Texts VII Qu. 15, VIII, XI), the micro-mechanical interaction of the two substances is then transmitted into the brain through the optic nerves and the continuity of their fibres, so that on reaching the exterior right and left sides of the *sensorium*, the motions of the two substances excite micro-mechanical motions of a sensitive substance in its pores.

Sixth (Texts VII Qu. 12, VIII) unlike the light-sensitive substance in the retinal pores, the sensitive substance in the pores of the *sensorium* is adapted to receive stimuli from *all* the external organs of sense; and, hence, the *sensorium* is the seat[94] of what Kepler had called '*sensus communis*'[95] and what Newton calls 'sense' or 'sensation'.

Seventh, because sensory perception is initiated by an external cause, it is a passion, not an action.

Before moving on to the motor system, I should like to draw attention here to the problem of applying modern terms anachronistically. So far, I have done this in the Introduction, as well as in this and the previous sections by applying the term 'receptor' and, in the above summary, the term 'end-organ' to the retina and the *sensorium*, both of which are organs or differentiated parts of an organism adapted for certain functions. Is this application appropriate? The answer is a qualified 'yes', as follows.

To begin with, the modern term 'receptor' means any organ or structure which, in receiving stimuli of a certain kind from the environment, generates nerve impulses that convey information regarding that environment.[96] In Newton's scheme for visual sensation, both the retina and the *sensorium* receive stimuli, the one directly from the environment (light rays), the other indirectly, so that the term 'receptor' could be applied to both organs. Nevertheless, when I occasionally use the term 'receptor', it must be remembered that in Newton's container/contained model of the nervous system, there is a dualism between the containers and the contained. And it is the contained sensitive substances that are Newton's true receptors.

As for the modern term 'end-organ', this means an organ or structure in which nerve fibre terminates.[97] As previously indicated, Newton's description of the anatomical pathway for vision proceeds from the fibres that exit behind the retina,

[94] OED has no entry for the special use in anatomy of the term 'seat'; but see various related Latin terms: *sedes* = seat (dwelling place), *situs* = site; and *locus* = place in which something is situated.

[95] See p. 33 above.

[96] See the supplement to OED 3b, *Biology*. The term was introduced as a term of art in neurology following its coinage in 1906 by Charles Sherrington.

[97] The main, as well as supplementary volumes of OED do not include this term; but see *Blakiston's Pocket Medical Dictionary* (4th edn., New York, 1979).

continue through the X-shaped structure of the optic nerves into the brain until the fibres terminate on the exterior of the right and left sides of the *sensorium*. Yet, as will become evident in Part III, the optic nerves originate in the brain and, thus, their fibres also terminate in the retina. Consequently, both the retina and the *sensorium* meet the modern definition of end-organ, the difference being that the retina is the peripheral and the *sensorium*, the central end-organ for vision.

Turning now to the motor system, it is important to note that although Newton's texts contain brief allusions to animal motion as an involuntary action,[98] yet his paradigm is voluntary action.

First (Text II, comment to Text IX), the internal cause of voluntary action is an immoveable principle called 'the soul' (*animus*), the seat of which is an undisclosed place in the anatomical space of the brain.

Second (Text II), the material cause of voluntary action is part of a corporeal, moveable principle (*anima*) called 'ætherial animal spirits'.

Third, (Text II), by an undisclosed chemical process, the ætherial animal spirits are perfected in the brain substance from ætherial vital spirits 'raised' during a chemical reaction in the heart.

Fourth (Texts II, X), only the ætherial animal spirits, not the soul, are extended throughout the pores of the containers in the nervous system.

Fifth (Texts II, X, XI), for voluntary action the soul excites, by impression, micro-mechanical motions of the ætherial animal spirits, as well as causally necessitates the anatomical pathway through which these motions are transmitted from the brain through the appropriate nerves and their fibres to the muscles.

Sixth (Texts II, X), to contract and relax a muscle, a small portion of ætherial animal spirits penetrates and then recedes from the fibres in the muscle, so that on contraction, the muscle increases its volume ('swells') and on relaxation, it decreases its volume ('shrinks').

Seventh, since contraction is initiated by an internal cause, animal motion is an action, not a passion; and since the soul (*animus*) determines the effect of animal motion, this immoveable entity is the agent or efficient cause of all animal motion that is a voluntary action.

To the above summaries may be added comments of an historical nature, beginning with Newton's investigation of optical processes. As indicated in the previous section, this was based on the new and correct physiological theory of Kepler, who solved the problem of how the eye formed an image[a] on the retina by isolating the geometrical optics of the eye from all other aspects of vision. In doing so, as Alistair Crombie and others have pointed out, he could treat 'the living eye on the model of a camera obscura with a lens' (crystalline humour) and also reform 'the optical geometry of how this device [the lens] focussed the picture on to the

---

[98] See *supra* Pt.I.1.2, Text II (regarding the heart), Text X (regarding palsy) and Texts IX, XI, XII (regarding the 'instinct' of animals'). Note that in the manuscripts, 'De motu et sensatione Animalium' and 'De vita & morte vegetabili', Newton retains the paradigm of voluntary action.

retinal screen'.[99] As a consequence, he could demonstrate that the retina is the peripheral end-organ for vision.

Nevertheless, the physiological explanations of both Kepler and Newton retain the basic outlines of Aristotle's theory in *De Anima*, III.i–iii.[100] As summarised by H. M. Koelbing, this theory holds that

> ...in vision as in every sensory perception the sense organ is always and exclusively acted upon from the outside. This *action*—originating in the visible object and transmitted to the eye by means of a transparent medium (air or water)—is described as a *movement*. In the eye, it causes an *alteration* of a particular nature which is finally perceived by the mind. Vision, then, is a one-directional process.[101]

But the two men differed from Aristotle in at least one important respect. For they placed the seat of sensory perception in a part of the brain that Newton calls 'the *sensorium*' but which Kepler identified only as the 'seat' of common sense. By contrast, Aristotle, in *Parva naturalia*, placed the seat of sensation in the heart, because, as he explained, 'in this part must lie the common *sensorium* of all the [external] sense-organs'. For 'if in all creatures life resides in this part, clearly so too must the origin of sensation; for we say that a creature is alive in so far as it is an animal, and an animal in so far as it is sensitive.'[102] Consequently, he regarded the heart not only as the starting point of life and movement, but also as the seat of the common sense faculty and the place of 'that sense organ which is master of all the rest',[103] namely, the common *sensorium*.

Kepler's experimental demonstration provided subsequent investigators not only with a scientific method, but also with a scientific basis for reintroducing the investigation of the complexities of sensory, as well as motor functions, yet those investigations would inevitably involve a review and critique of older theories.[104] Indeed, as indicated in the previous section, the Galenic tradition is evident in Newton's account of the X-shaped structure of optic nerves, as well as in his brief

---

[99] Crombie, *Science, Optics and Music*, pp. 170–1.

[100] See Hett (tr.), *Aristotle On the Soul*, pp. 140–63, and Lawson-Tancred (tr.), *Aristotle De Anima*, pp. 189–201.

[101] Koelbing, 'Ocular Physiology', p. 219; who also pointed out, pp. 219–20, that the notion of vision as a one-directional process (action of an object, movement of a medium, alteration in the eye) was not generally accepted either by philosophers such as Plato or by the Greek naturalists, including Galen, whose physiological theory, which was similar to Plato's '*synaugeia*', became known as the emission-intromission theory.

[102] See Hett (tr.), *Aristotle on the Soul*, 'Parva naturalia', pp. 418–21, p. 421.

[103] Ibid., pp. 326–9, p. 329. On the common sense faculty, see also Tracy, *Physiological Theory ad the Doctrine of the* Mean, pp. 212–20, as well as p. 34 above. On Aristotle's use of the term 'faculty' (*dunamis*), see Lawson-Tancred in *Aristotle De anima*, p. 118; and Aristotle's conception of the brain as a refrigerating agent, see Woollam, 'Concepts of the Brain and its Function in Classical Antiquity', p. 11.

[104] For example, Harvey, in a number of his writings, critically reviewed the classical sources on various topics (e.g., generation, movement), as well as attempted to harmonise the radical contradictions between his own opinions and those of Aristotle, as well as the Galenic physicians; for two instances of this, see Kassler, *Music, Science, Philosophy*, pp. 61–82, 98–100.

critique of three Galenic hypotheses of muscle action, to which he responds with his own Galenic variant. But Newton also is indebted to another, very important feature of Galenic tradition, for, as Thomas Hall has pointed out, Galen's anatomical system helped to establish a standard method of explaining physiological function, because it offered 'a kind of master strategy of interpretation that was to influence later biomedical thinking right up to the twentieth century'.

According to Hall, in this Galenic tradition of interpretation,

> ...physiological explanation entails two analytical steps. First, sensible activities of the body are restated in terms of insensible activities (usually micromovements of some sort) which are viewed as *latent equivalents of patent manifestations* of life. This part of the method had been in use since pre-Socratic times. Second, these latent equivalents are attributed to something more fundamental—to a disposing principle, property, or power that permits the micromovement in question. Such principles, properties, or powers have been viewed as legitimate tools of explanation even when left unexplained or even when presumed to be inexplicable in themselves.[105]

In other words, disposing principles, properties or powers are *unknown causes*, as may be illustrated briefly with reference to the sixteenth-century physician, Jean Fernel.

In his influential treatise, *De naturali parte medicine*,[106] Fernel followed the physiology, as well as the interpretive strategy of Galen. But he rejected Galen's derivation of the soul from the blend of the body. Instead, his term 'soul' is a term of convenience for an unknown cause, knowledge of which is 'exceedingly involved and obscure and, as with whatever is not obvious, is not at all easily attained'. Indeed, Fernel described the soul as 'a fugitive from sensation', because our mind discerns it only through the functions and works it brings forth'. And he therefore claimed that, similarly, 'we are wont to discover *all hidden causes whatsoever* ... by those of their effects', which are 'conspicuous and ... present themselves to sensation'.[107] In short, as the last term in Fernel's causal-analytical system, the soul, which in Galenic fashion he treated as the irreducible unknown cause of all functions of the body, can be known exclusively only through its effects.

The causal-analytical system of Newton also involves distinguishing between effects (i.e., products or work), actions (i.e., micro-mechanical motions of unobservable corporeal substances productive of effects) and unknown causes that include, but are not restricted to the disposing principles of *animus* and *anima*. It is ironic, therefore, that from the eighteenth century, a number of physiologists modelled, or claim to have modelled their procedures not on those of Galen but on

---

[105] Hall, 'On Biological Analogs', p. 10, who also indicated, p. 11 that, after Galen, dispositional thinking was furthered by the Scholastics, who re-emphasised a distinction, traceable to Greek Stoicism, between 'proegumenal' (or dispositive) and procatarctic' (or incitative) causes. In Newton's day, the latter cause was understood medically as an external and occasional cause such as the previous disposition of a patient; see *supra* Pt.I.1.2, comment to Text IV.

[106] The first edition appeared in 1542, 'within about a year of the publication of the great works of Copernicus and Vesalius'; see Hall, 'On Biological Analogs', p. 10. According to Brazier, 'The Historical Development of Neurophysiology', p. 2, this was the first textbook on physiology.

[107] Quoted in Hall, 'On Biological Analogs', p. 10 (italics mine).

those of Newton,[108] their main reference point being not only the latter's treatment of the law of universal gravitation,[109] but also his agnosticism about causes.[110]

Galen's physiology, like Aristotle's, was teleological, as well as vitalist.[111] For having anatomically divided the body into separate systems of the liver, heart and brain,[112] he assigned vital functions to the production and micro-movements of a vital principle conceived as a hierarchy of three spirits (*pneumata*): the liver's natural spirit (carried by the veins), the heart's vital spirit (carried by the blood) and the brain's animal spirit (carried by the nerves). As Arthur Brock has pointed out, the lowest of Galen's three spirits, natural spirit, presided over subconscious vegetable life and, hence, was nearly equivalent to nature.[113]

In Newton's day, William Harvey rejected spirits, whereas Descartes reduced their number to animal spirits only.[114] Nevertheless, there were a number of attempts to retain a non-mechanical vital principle as a means of overturning Descartes's mechanistic physiology.[115] For example, the Cambridge Platonist, Henry More, reintroduced Galen's natural spirit as a 'hylarchic' spirit of nature that rules over matter and described this spirit as follows:

> ...everything that is called body I hold to be alive in a sottish and drunken sort of way, inasmuch as it [the spirit of nature] is an image and the lowest and basest shaddow [*sic*], though destitute of [the image of the higher shadows,] sense and animadversion, of the Divine essence'.[116]

As my second interpolation suggests, More conceived his vital principle as a hierarchy of three spirits, that, in ascending order, are called 'nature', 'sense' and 'animadversion' (cognitive perception). Moreover, he further conceived these three

[108] See ibid., pp. 12–4, who did not mention Robert Whytt, one of the pioneers in the study of reflex action, who adopted physiological procedures in the Newtonian style at the same time retaining the Galenic tradition of treating unknown causes, which in his case include a disposing 'sentient' principle co-extensive with 'mind'; see Boring, *A History of Experimental Psychology*, p. 35; see also Clarke and Jaycna, *Nineteenth-Century Origins of Neuroscientific Concepts*, pp. 113, 402 n.83.

[109] As is well known, Newton sought, but failed to discover the unknown causes of gravity, cohesion and fermentation, so that these physical phenomena are instances of disposing 'powers'.

[110] For an instance of his agnosticism about causes in relation to a disposing 'property', see *supra* Pt.I.1.2, comment to Text IV.

[111] See Wightman, *The Emergence of Scientific Medicine*, pp. 44, 91. For the medical background to, and details of Aristotle's physiological concepts, see Tracy, *Physiological Theory and the Doctrine of the Mean*, pp. 197–222.

[112] For the brain system, see Spiegel, *Galen on Psychology*; see also Woollam, 'Concepts of the Brain and its Functions in Classical Antiquity', pp. 17–9.

[113] See Brock (tr.), *Galen on the Natural Faculties*', pp. xxxiv–xxxv, who, p. xxv, described Galen's book as providing instances within 'its comparatively short compass' that illustrate 'perhaps most of the sides of this many sided writer'. For a succinct but useful summary of Galen's physiology, see also Crombie, *Medieval and Early Modern Science*, vol. 1, pp. 162–8.

[114] For Harvey, see *infra* Pt.III.3.2, p. 82; for Descartes, see p. 42 above.

[115] See *supra* Introduction, p. viii and n.14

[116] Quoted in Gabbey, 'Henry More and the Limits of Mechanism', pp. 27–8. See Kassler, *Seeking Truth*, pp. 96–7 and n.182; see also Jacob, 'The Neoplatonic Conception of Nature', pp. 105–10.

spirits as mere worldly appearances, images or shadows of a reality called 'the Divine essence'. In this kind of Platonising, therefore, More followed pre-Cartesian Renaissance pneumatology, when, according to Thomas Hall, three spirits were the rule.[117]

However, some of More's contemporaries did not follow this rule, even if they too were concerned about Cartesian mechanism. For example, Willis conceived his vital principle as a hierarchy of two spirits, vital and animal,[118] just as Newton does in Text II. It is important to note, therefore, that between the date of Text II and the publication of the first 1687 edition of *Principia mathematica*, Newton's procedure had changed, for instead of setting up two special vital powers, he reduced 'vital action to two irreducible chemical powers of attraction and repulsion'.[119] Regardless of the change, both kinds of powers (vital, as well as chemical) are what Hall referred to above as '*latent equivalents of patent manifestations of life*', namely, the micro-mechanical motion of unobservable corporeal substances that produce effects, although the cause of the latent equivalents is unknown.

Galen was an eclectic: Hippocratic as a physician, Platonic as a philosopher and Sophistic as a rhetorician.[120] Although he often followed Aristotelian precedent, he rejected Aristotle's doctrine of the heart as the central organ of the body and as a container in which innate heat is produced and distributed with the blood. Instead, Galen identified the heart as the seat of the non-rational (affective) part of the soul and the brain as the central organ of the body and its substance as the seat of the rational (reasoning) part of the soul.[121] He thus combined Aristotle's principle of animal life (*anima*)[122] with Plato's principle of intellectual life (*animus*).[123] And, as indicated in the previous section and in the summaries above, this kind of dualism is present in some of Newton's texts.

Finally, although 1673 seems to be the earliest date for Newton's *public* mention of the human *sensorium*,[124] yet it is highly probable that his interest in this entity was stimulated by reading a book entitled, *The Immortality of the Soul*. Published in 1659, its author, the previously mentioned Henry More, included a critical survey of eleven opinions concerning the anatomical seat of the faculty called 'common

[117] See Hall in Descartes, *Treatise of Man*, p. xxviii.

[118] See pp. 42–3 above.

[119] Hall, 'On Biological Analogs', p. 13.

[120] For an instance of his eclecticism, see Galen, *On the Doctrines of Hippocrates and Plato*, which includes vigorous criticisms of the Aristotelian Stoics and more tempered criticisms of the Platonic Stoic, Posidonius (see, e.g., pp. 329, 331). Perhaps, therefore, the latter was the source for Stoic influences noted by Brock, 'Introduction', *Galen on the Natural Faculties*, p. xxv.

[121] See, e.g., Galen, *On the Doctrines of Hippocrates and Plato*, pp. 157, 211, *et passim*. See also Woollam, 'Concepts of the Brain and its Functions in Classical Antiquity', pp. 13–8.

[122] See p. 51 above.

[123] For the medical background to, and details of Plato's *Timaeus*, see Tracy, *Physiological Theory and the Doctrine of the Mean*, pp. 77–156.

[124] See *supra* Pt.I.1.2, Text I.

sense'.[125] Indeed, Newton not only owned a copy of the first edition,[126] but also, as a student, made notes on this very topic, probably from the copy in Barrow's library.[127] The eleven opinions surveyed by More assign the seat of common sense as follows:

Erasistratus of Alexandria (via Galen): in the membranes;
Plotinus and Thomas Aquinas: in the whole body;
J. B. van Helmont: in the stomach;
Aristotle and Thomas Hobbes: in the heart;
Plato: in the brain;
Kenelm Digby: in the so-called 'septum lucidum' that divided two of the ventricles in the brain;
Descartes: in the pineal gland;
Thomas Wharton: in the spinal marrow or nerves about the fourth ventricle; and.
Henricus Regius: in a small, perfectly solid particle within the body.[128]

After criticising each of these opinions, More offered his own opinion, namely, that the common sense is located in the fourth ventricle of the brain, thus repeating the opinion of Herophilus of Alexandria (whom he did not name).[129] And the reason he gave for his opinion was that

> ...that Matter which is *the immediate Instrument* of all the Animal functions of the Soul, should be the chiefest Seat from whence and where she[130] exercises these functions, and if there be any place where there is freer plenty of the purest sort of this Matter, that [place should be] her peculiar residence.... Now the *immediate Instrument* of the functions of the Soul is that thinner Matter which they ordinarily call *Animal Spirits* which are to be found in their greatest purity and plenty in the *fourth Ventricle of the Brain*. From whence it must follow that that precious and choice part of the Soul which we call *the Centre of* [sensory] *perception* is to be placed in that Ventricle ... in the midst of these Spirits themselves; for that [ventricle] is the most natural situation for commanding them into the parts of the *Head* and *Body*; besides a more delicate and subtile use of them at home, in pursuing various imaginations and inventions.[131]

---

125 See More, *Philosophical Writings*, pp. 129–49; see also the useful notes by the editor, pp. 321–4.

126 See Harrison, *The Library of Isaac Newton*, (No. 1113), p. 196.

127 See McGuire and Tamny (eds.), *Certain Philosophical Questions*, p. 183; see also Feingold, 'Isaac Barrow's Library', (No. 663), p. 358.

128 For details, see MacKinnon in More, *Philosophical Writings*, 'Notes', pp. 321–9.

129 According to MacKinnon in ibid., p. 323, Galen is the probable source for More's own opinion. Note, therefore, that it was Herophilus, who gave the first clear description of the ventricles; see Woollam, 'Concepts of the Brain and its Functions in Classical Antiquity', p. 12. Note also that after him, thought about the brain concentrated on the ventricles for nearly two thousand years; indeed, according to Clarke, 'Brain Anatomy before Steno', p. 30, in 'all descriptions of the brain' after Herophilus, the anatomist 'paid more attention to the ventricular system than to any other part because of its functional significance'.

130 More, here and elsewhere, followed the traditional distinction between God as masculine and the human soul as feminine; but as will later become apparent, some of those who adumbrated a dual soul concept, described the inferior soul alone in feminine terms.

131 More, *Philosophical Writings*, p. 143.

By the 'choice part' of the soul, More denoted animal spirits of the 'greatest purity', and these are the faculty which he called either 'common sense' or 'common percipiency'. As for his reference to a '*Centre of perception*', which he also named the 'Common *Sensorium*',[132] More did not echo Aristotle, who described that centre as an organ; rather, he echoed Aristotle's followers, who referred sensory perception to a common centre of perception. Accordingly, More assigned the purest and most plentiful animal spirits to the fourth ventricle, thereby ruling out the 'middle' ventricle, because its spirits were 'not so pure', as well as the 'upper' two ventricles, which 'being two', were 'not so fit for the office' to be performed by the animal spirits, namely, '*Sense* and *Motion*'.[133]

But he did not mention two other reasons that would also explain why he settled on the fourth ventricle not only as the seat of the soul, but also as the centre of perception and its common-sense faculty, the animal spirits. For, on the one hand, in his book, *An Antidote against Atheism*, published in 1652, he propounded Plato's doctrine of the soul as a recollection.[134] According to this doctrine, knowledge is not acquired through the senses or as information conveyed from one mind to another by teaching; rather it is acquired by recollection in this life of truths seen and unknown by the soul before its incarnation. And, on the other hand, during the long period in which psychic faculties (*pneumata* or animal spirits) were localised in the ventricles of the brain, it was the opinion of many writers that the fourth ventricle was the seat of memory.[135] However, by 1616, when William Harvey began to prepare notes for his Lumleian lectures on anatomy, he simply recorded the latter opinion as past history by citing the names of three medieval authors as instances of 'those who dispose the faculties according to the ventricles ... [and who] locate memory here', that is, in the fourth ventricle.[136]

Of course, in retaining the fourth ventricle as the seat of the soul and centre of perception, More's purposes were theological, not anatomical. For from that seat, he wrote, the immaterial soul '*perceives all Objects*, ... [as well as] *imagines, reasons and invents, and from whence she commands all the parts of the body*'.[137] But she does not move away from her seat, because her commands are carried out

[132] See OED, where the first citation in English of the term, *sensorium*, is to a 1647 poem by More. See also MacKinnon in More, *Philosophical Writings*, p. 291–2.

[133] More, *Philosophical Writings*, pp. 144–5.

[134] See Kassler, *Inner Music*, pp. 115–7.

[135] See Woollam, 'Concepts of the Brain and its Function in Classical Antiquity', and Pagel, 'Medieval and Renaissance Contributions to Knowledge of the Brain and its Functions'. In his book, *Gargantua et Pantagruel*, the sixteenth-century physician, François Rabelais, using the body as an extended metaphor, foreshadowed the demise of the ventricular theory by giving a satirical account of it, whereas after the anatomical work of Vesalius, the theory was no longer tenable.

[136] See Harvey, *Lectures on the Whole of Anatomy*, p. 222, which forms part of his notes on the brain and nerves, pp. 215–227. These notes, which were not published during Harvey's lifetime, survey anatomical opinions then in process of re-assessment and criticism. For his mature 1651 opinion concerning brain function, in which he compared conception in the brain with conception in the womb, see Kassler, *Music, Science, Philosophy*, pp. 61–82.

[137] More, *Philosophical Writings*, p. 141.

by the animal spirits flowing out of the fourth ventricle through the substance of the brain into nerves. Note, however, that More conceived the brain as 'one great round *Nerve*', which, like the nerves that are 'so many smaller productions or slenderer prolongations of the *Brain*', have 'invisible porosities'. Consequently, he regarded the porous substance of the brain and its prolongations as a 'continued Receptacle of that immediate Instrument of the sensiferous motions of the Soul, the *Animal Spirits*, wherein also lies her hidden Vehicle of Life in this mortal body'.[138]

So much, then, for this historical digression, which is included primarily as a reminder that Newton's texts contain a mixture of old and new concepts.[139] As for the conclusions that precede this digression, these present in a summary form what can be discovered from Newton's texts, as presented in Part I, concerning sensory-motor function in relation to the human *sensorium*. Unfortunately, however, they also point to gaps in our knowledge, the most important of which may be posed in the form of four questions as follows:

(1) Since Newton treats as receptors both the light-sensitive substance in the retina of the eye and the sensitive substance in the *sensorium* of the brain, does this mean that he also extended to the *sensorium* the *camera obscura* model that he used in Text III to explain not only the functional anatomy of the eye, but also image formation on the retina?
(2) Since the *sensorium* is the seat of sensory perception, is it also the seat of locomotion?
(3) Where in the anatomical space of the brain is the place of the *sensorium*?
(4) And where in the brain is the seat of the immovable soul (*animus*) that acts on its corporeal and movable instrument (*anima*)?

These questions cannot be answered, even conditionally, from Newton's texts intended for a public audience. Consequently to seek answers to all four questions, recourse must be had to a wider context, which in Part III is represented by certain anatomical and natural-philosophical concepts of the physicians, Briggs and Willis. The purpose of Part III, therefore, is to examine some of those concepts insofar as they relate to this inquiry, beginning with Briggs, concerning whom it will be necessary at the outset to provide some biographical particulars in order to determine when he and Newton first became acquainted.

[138] Ibid., p. 145.

[139] See Hall in Descartes, *Treatise of Man*, pp. xxxi–xxxiii, who pointed out first, that although Descartes's physiology was derivative of traditional ideas, yet he discarded certain of them in order to build a new mechanistic physiology of reflex; and second, that of his predecessors 'it will be Galen whose name and whose ideas will appear most frequently in our notes'.

Part III

III
The Human *Sensorium* in Wider Context

3.1 William Briggs on the Visual Sensory System

In 1663, two years after Newton was admitted to Trinity College, Cambridge, Briggs was admitted to Corpus Christi, graduating B.A. (1667) and M.A. (1670, incorporated at Oxford).[1] In 1668 he became a fellow of his college, the same year that Newton became a major fellow at Trinity. The patronage of Ralph Montagu, who on 1 January 1669 was appointed ambassador-extraordinary to France, enabled Briggs to travel there for further study,[2] which included attending, in Montpellier, the anatomical lectures of Raymond Vieussens. According to Briggs, these lectures extended his learning beyond 'the elegant structure of the parts of the eye and its labyrinthine convolutions' to the whole of the 'animal economy', including the thin films or 'overall threads of the human body'.[3]

Although the date of his return to Corpus Christi is unrecorded, it must have been sometime in 1672, when in May of that year Charles II terminated Montagu's ambassadorship. Not long afterwards Briggs commenced writing his treatise, *Ophthalamo-graphia sive oculi ejusq; partium descriptio anatomica*; and after its publication in 1676,[4] he was awarded an M.D. (1677). In 1682, having been appointed as physician to St. Thomas Hospital, he resigned his fellowship and moved to London, publishing there in English two short tracts on vision, one in March 1682, the other in May 1683.[5] Sometime afterwards, Newton encouraged

[1] His tutor was Thomas Tenison, later Archbishop of Canterbury, who was one of the patrons of Samuel Clarke, the translator of Newton's *Optice*; see Kassler, *Seeking Truth*, p. 116 *passim*.

[2] See Briggs, *Opththalamo-graphia* [1990], 'Epistola Dedicatoria' to Ralph Montagu.

[3] Ibid., 'Præfatio', which contains the only mention of Vieussens in Briggs's writings, which suggests that his conception of the human body's thin films, i.e., fibres or filaments, was afterwards superseded by his exposure to Willis's anatomy.

[4] Barrow owned a copy of this edition; see Feingold, 'Isaac Barrow's Library', (No. 191), p. 346.

[5] Briggs, 'A New Theory of Vision', and 'A Continuation'.

J. C. Kassler, *Newton's* Sensorium*: Anatomy of a Concept*, Archimedes 53,
https://doi.org/10.1007/978-3-319-72053-1_3

Briggs to republish his 1676 Latin treatise and to include, in a Latin dress, the two tracts.

This 'Editio altera', published in 1685,[6] had two parts, each separately paginated and with separate title pages, one for the re-issued Latin treatise and one for the two tracts translated into Latin.[7] Immediately following the second title page is a commendatory letter written by Newton,[8] in which he recalls that Briggs's 'expert dexterity' in dissecting an eye was 'once no small delight to me'. On that single occasion,

> You laid out the [eye's] motor muscles in their natural setting and exposed the other part openly so that one could not so much understand as clearly discern their individual uses, with the result that for some time I have expected great precision from your knife. Nor was my hope betrayed by that excellent anatomical treatise *which you later published*.[9]

The words italicised above provide evidence that Newton and Briggs became acquainted prior to publication of the latter's 1676 treatise. And more evidence is found in the treatise itself, where Briggs describes Newton as 'my learned friend' and afterwards mentions briefly the latter's 'hypothesis' concerning the colours of thin films of natural bodies such as silk and the iris of the eye.[10] Briggs himself proposed an hypothesis concerning the colours of the latter,[11] which he attributed to the varying mixture and distribution of heat and spirit coming from the 'lobes' of the brain, as well as to the subtle texture of the 'filaments' of the iris itself. But he

[6] Newton owned a copy of this edition; see Harrison, *The Library of Isaac Newton*, (No. 297) p. 111.

[7] The 'Editio altera' of 1685 was reissued at Leyden in 1686. A facsimile of the 1685 edition was published in 1990 with an English translation of all three publications; see Briggs, *Opthalamo-graphia* [1685/1990, part 1]. Unfortunately, there are a number of problems with the translation, including the literal rendering of some seventeenth-century technical terms that can be misleading (for some other problems, see n.8 and p. 67 n.39 below). In addition, the editor seems to have substituted modern terms that are inappropriate for a seventeenth-century text (e.g., '*chiasma*' for 'unionem', 'ciliary body' for 'processus ciliare' and, in one place 'oxygen' for 'spirituum'). Consequently, when quoting from the English translation of the republished 1676 treatise, I have checked the translation against the original version. As for Briggs's two tracts, I rely solely on the original English texts of 1682 and 1683.

[8] In the 'Editio altera' of 1685, the letter is dated 'Cantabrigiæ 7 Kal. Maii, 1685'; but in the English translation of this letter in Briggs, *Opthalamo-graphia* [1990], the letter is dated 25 April 1685; and this same discrepancy also appears in Newton, *The Correspondence*, vol. 2, Letter 280, pp. 417–8 (Latin version), 418–9 (English translation).

[9] Briggs, *Opthalamo-graphia* [1990], p. v [italics mine].

[10] See Briggs, *Ophthalamo-graphia* [1676], pp. 17, 21.

[11] Ibid., pp. 17–22, who also claimed that the texture of the filaments of the iris was evident by analogy with the phenomena of other natural bodies, including feathers of birds, silk and flowers.

declined to indicate whether his hypothesis agreed with those of the 'distinguished' Descartes or the 'ingenious' Newton.[12]

Newton, however, did not include the iris in his mature explanation of the colours of thin films, which forms part of Book II of *Opticks (1)*.[13] Neither did he mention the iris in the earlier explanation found in 'An Hypothesis' (the source of Text II), where, in the concluding part of the paper read to the Fellows of the Royal Society on 10 February 1667, he considered 'how the phænomena of thin transparent plates stand related to those of all other natural bodies'.[14] But even as early as 1666 he was already at work on some of the ideas adopted in 'An Hypothesis'; and about 1672 he had developed a concept to explain the colours of thin films.[15] It therefore is probable, first, that Newton and Briggs became acquainted shortly after 1672, when the latter left the continent and returned to Cambridge; and, second, that between 1673 and 1675 Briggs learned about the colours of thin films, though, perhaps, without fully understanding Newton's explanation of them.[16]

In his 1676 treatise, which Newton describes as 'outstanding',[17] Briggs, without mentioning the *camera obscura*, presented an account of optical processes in which the retina serves as the 'primary' organ of sight.[18] However, because of the varying refractions of the light rays, as well as the various 'consistency' of the aqueous, crystalline and vitreous humours, he conceived all three humours as refracting lenses, thereby modifying Kepler's functional anatomy of the eye.[19] But in defending

---

[12] Ibid., p. 21. Of the three works by Descartes (*La Dioptrique*, *L'Homme*, *Les Météors*) cited by Briggs, it is probable that his reference above is to the corpuscular explanation of the colours of the rainbow and clouds in the last work and which concluded Descartes, *Discourse on Method*, pp. 332 *seq*.

[13] But he did mention the colours of thin films such as 'the fine Hairs, or *Capillamenta*' of bird feathers, silk, cloths and flowers; see Newton, *Opticks (1)*, Bk. II, Pt.II, Props. V–X, pp. 55–6. He also mentioned 'the Webs of some Spiders', which, in terms of thin films, implies webs called 'gossamers' that are spun by small spiders.

[14] Cohen and Schofield (eds.), *Isaac Newton's Papers*, pp. 226–38, in which, pp. 229–30, mention is made of silk, cloths and flowers.

[15] The concept in question is Newton's 'fits' of easy reflection and easy transmission; see *supra* Pt.I.1.2, Text IV. But in 'An Hypothesis', his explanation is in terms of mechanical æther vibrations and not, as in *Opticks (1)*, in terms of attractive powers.

[16] See Shapiro, *Fits, Passions and Paroxysms*, pp. 77–9, where, on p. 79, he mentions the assumption Newton had to make when applying his hypothesis of aetherial vibrations to the periodic colours of thin films, namely, that the æther vibrations stirred up by the incident light rays move faster than the rays themselves; see also *supra* Pt.I.1.2, comment to Text IV. Compare this assumption with Briggs, *Ophthalamo-graphia* [1676], pp. 54–5, who mentions a '*materia ætheria*' when indicating that under certain circumstances, the vibrations of the light rays exceed the motion of the animal spirits.

[17] See Newton's commendatory letter in Briggs, *Ophthalamo-graphia* [1990 part 2].

[18] I.e., the peripheral end-organ of vision; see Briggs, *Ophthalamo-graphia* [1676], p. 17.

[19] Ibid., p. 25. This modification may have been due to Scheiner, whose work is cited, p. 36, but not in connection with humours as refracting lenses.

the function that Kepler had assigned to the retina, Briggs's offered two arguments. On the one hand, since the retina shares the same substance as that of the brain, it is 'suitable for retaining and passing on all the impressions of objects'. On the other hand, because 'the lucid rays' penetrate deeply into the fibres of the retina, those rays affect the animal spirits contained in the pores of the retina, where the rays and the spirits are 'as if mixed'.[20]

From these statements we may conclude, first, that the substance under the coat of the retina is the same as that under the coats of the brain; second, that the retina contains a receptor in the form of a light-sensitive substance called 'animal spirits'; and third, that when the rays mix with this substance, the mixture constitutes the retentive part of the optical process. What, then, constitutes the transmitting part? Briggs's answer is that the optic nerves 'communicate the impressions of the rays to the middle of the brain'.[21] For these nerves, which

> ...come off from close to the corpora striata[,] ... are of an exceptional size and are bestowed with innumerable fibers ... so that they in turn may be enlarged and a more copious store of *spiritus* may be supplied to the eye in vision. ... [As for the nerves themselves, they] consist of two tunicae [i.e., coats] which proceed from both meninges[22] of the brain ... and a medulla[23] or fibers, which descend in an ordered series to the fundus[24] of the eye and sclera.[25] While we acknowledge that these medullar[26] fibers escape our vision at the union [of the optic nerves], we suppose ... that they are carried down from the brain in a separate fashion....[27]

Two points in the above passage deserve comment. First, its initial clause contributes some new information to our inquiry. For if the optic nerves originate near two bodies called 'corpora striata', then those two bodies, as well as the place where the optic nerves originate, must be located somewhere in the middle of the brain. Although there is no further mention of the term 'corpora striata' in Briggs's three publications, yet an illustration in one of his English tracts provides some insight into the location of these two bodies in the brain, as will soon become evident.

The second point concerns Briggs's supposition that the fibres descend from the brain in a separate fashion, even though they cannot be seen at the union of the optic nerves. But does the term 'union' denote an X-shaped decussation or crossing of the

[20] Briggs, *Ophthalamo-graphia* [1990], pp. 17–8, 35.

[21] Ibid., p. 17.

[22] I.e. the brain's membranous coverings called the dura and pia mater.

[23] I.e., center (of the nerves).

[24] I.e., (from the exterior of the eye inward) the region of the vitreous humour in front of the retina; see Briggs, *Ophthalamo-graphia* [1990], pp. 24–5.

[25] I.e., (from the exterior of the eye inward) the outermost coat of the eye, of which the cornea is the part in front of the aqueous humour that precedes the crystalline humour.

[26] I.e., marrowy or medullary, not membranous.

[27] Briggs, *Opththalamo-graphia* [1676], pp. 42–2; [1990], p. 28.

nerves and their contained fibres? This question cannot be answered from the passage quoted above. Nor can it be answered from another passage concerning the position of the eyes,[28] in which Briggs briefly addressed the problem of binocular fusion—why only a single sensation arises from two eyes. According to his explanation, the two eyes are suitably balanced

> ...so that the vibrations of the rays from a single object are transmitted to the brain through filaments of the optick nerves which are *homologous* (namely they correspond exactly to each other); and thus our mind (as when two citharae of a unisonous chord are struck at once) perceives only one movement.[29]

Fortunately, in the short tract of March 1682, Briggs made his understanding of the term 'union' very clear:

> ...that the [optic] Nerves *decussate* [i.e., cross] one another, or that they are blended together so in their Union as to cause a confusion in their Fibres, is not to be imagined; but those [fibres] that are in the *thalami Optici* [i.e., optic chambers] on the right side run distinctly to the *right Eye*, and those on the left accordingly.[30]

In other words, the optic nerves, with their contained fibres, unite but do not decussate'.

Briggs's position may have been common among physicians, since during the more than forty years that Harvey delivered his Lumleian lectures on anatomy at the Royal College of Physicians in London, he used the notes prepared between 1616 and 1618 'again and again, ... adding others in the margins or on blank versos or even inserting odd leaves as occasion required'.[31] In one of his notes, he wrote:

> Some [declare that the fibers] of the optic nerves [are arranged] in [the form of] a cross, wherefore the object does not appear [as] 2 and the eyes are moved simultaneously. ... [There are many] opinions regarding the juncture which Vesalius refutes by three examples; for in some they are not [as in] *ratts*, nor in some men.[32]

Unlike Briggs, Newton offered no opinion concerning whether the fibres decussate[33]; indeed, from the writings he made public, it might be inferred that he was unaware of the issue.

Furthermore, unlike Briggs, he ignores the relationship between the optic nerves and the optic chambers,[34] whereas at the conclusion of his tract, Briggs described these latter parts as two '*gibbous*' (convex) protuberances in the brain. And he

---

[28] See ibid. pp. 2–4, in which Briggs seems to assume that visual spatial localisation is the result of some kind of position sense of the two eyes from the tensions in muscles extrinsic to the eyeball. For the basis of this assumption, see *supra* Pt.II.2.1, pp. 38–9.

[29] Ibid., p. 3.

[30] Briggs, 'A New Theory', p. 170.

[31] See the translators' 'Introduction' in Harvey, *Lectures on the Whole of Anatomy*, p. 10.

[32] See Harvey, *Lectures on the Whole of Anatomy*, p. 225.

[33] See *supra* Pt.II.2.1, p. 38.

[34] The term 'thalami' had been coined by Galen; for its place in his anatomy, see *supra* Pt.II.2.1, p. 36 n.24.

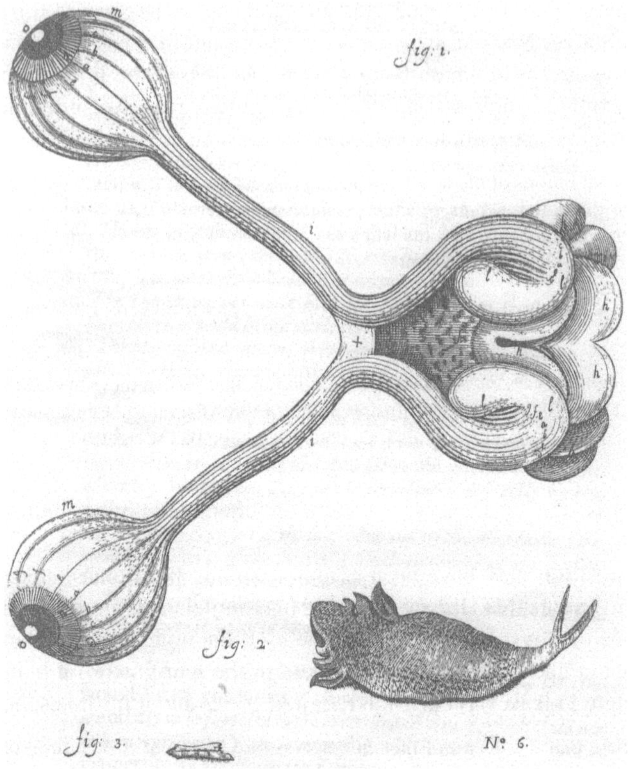

Fig. III.1 X-shaped structure of the optic nerves *from* Briggs, 'A New Theory of Vision' (1682), Fig. 1

hhhh	Parts of the brain that lie under the pair of optic chambers (thalami optice)
llll	Pair of optic chambers from whence the nerves have their rise
ii	Pair of optic nerves and their fibres before and after their union
+	The place of their union

accompanied this description with an illustration (see Fig. III.1).[35] According the explanatory table accompanying the illustration, the letter *l* marks the optic chambers '*from whence the Optick Nerves have their rise*'. Since in the passage quoted previously from his 1676 treatise, he had pointed out that the optic nerves 'come off from close to the corpora striata', it is possible that the parts called 'corpora striata' might be those in the illustration marked by the letter *h* and described in the explanatory table as the '*parts of the Brain that lie under*' the optic chambers. But Briggs's term '*under*' presents a problem, for in the illustration, the optic chambers are placed between the diverging portion of the parts marked with the letter *h*, which seem to lie *above* or *behind* the optic chambers. If this is the case, then it is possible

[35]This illustration faces the title page of No. 6 of *The Philosophical Collections*, whereas the explanatory table (Tab. I.) is found at the conclusion of Briggs, 'A New Theory', p. 178.

that *h* represents the two corpora striata and, hence, exhibits the right and left exterior sides of what Newton calls the '*sensorium*'.

Unfortunately, Briggs provided no additional information concerning the parts in the brain marked *h* in the illustration. But such anatomical details were subsidiary to the main purpose of his tract, which was to present a more developed explanation of the problem of singleness of vision (binocular fusion) first mooted in his 1676 treatise. In brief, his new theory is that

> ...the Fibres [in the optic nerves] that are parallel or in the *same position* are as it were *symphonical* ... and exactly answer one another in site and tension, so that when any impression from an object without moves *both Fibres*, it causes not a *double sensation* no more than *Unisons* in two *Viols* struck together cause a *double sound*.[36]

As for the optic chambers, he claimed that they assist in keeping the tension of forementioned fibres, because those chambers are analogous to the place where 'the strings of a Viol' are flexed 'upon the bridge of it'.[37]

Not long after the March 1682 tract appeared in print, Newton wrote to Briggs from Cambridge concerning the latter's 'very ingenious Theory of Vision', pointing out that

> ...some things [are] more solid & satisfactory, others more disputable but yet plausibly suggested & well deserving ye consideration of ye ingenious. The more satisfactory I take to be your asserting yt we see wth both eyes at once, your speculation about ye use of ye *musculus obliquus inferior*,[38] your assigning every fibre in ye optick nerve of one eye to have it's [*sic*] correspondent in yt of ye other, both wch make all things appear to both eyes in one & ye same place [i.e., situation] & your solving hereby ye duplicity of ye object in distorted eyes....

For Newton, the 'more disputable' suggestion

> ...seems your notion about every pair of fellow fibres being unisons to one another, discords to ye rest, and this consonance making ye object seen with two eyes appear but one for ye same reason that unison sounds seem but one sound.[39]

Then, in a second letter to Briggs, dated 12 September 1682, Newton communicated some criticisms concerning Briggs's analogy between 'audibles' and 'visibles',[40] part of which concerns a mechanical analogue of strings stretched over the bridge of a viol, so that 'the bending of ye nerves in ye Thalamus opticus should cause a differing tension of ye Fibres'. Instead, Newton resorts to an organic ana-

---

[36] Briggs, 'A New Theory', p. 169.

[37] Ibid.

[38] See ibid., pp. 170–1, where Briggs claimed that this muscle keeps the eye even and '*in sight*, so that it cannot incline too much either to the internal or external *Canthus*'. An errata list at the end of Briggs's tract corrects '*in sight*' to '*in situ*'. Unfortunately, in presenting the English version of the tract in *Opthalamo-graphia* [1985/1990, part 2], p. 5, the editor failed either to compare it with the translator's Latin version (which has 'in *æquilibrio*') or to consult the original 1682 errata.

[39] Newton, *The Correspondence*, vol. 2, Letter 261 (20 June 1682), pp. 377–8, with the address: 'For his honoured friend Dr. William Briggs at his house in Suffolk Street in London'.

[40] See ibid., vol. 2, Letter 264, pp. 381–5; for the criticisms, see pp. 381–3, 385.

logue for 'that most tender substance of ye Optick nerves wch grew bent from ye very beginning'.[41]

And this line of thought leads him to the problem of singleness of vision or, in his words, how 'the two motional pictures **in ye sensorium** come together & become coincident'.[42]

> If when we look but with one eye it be asked why objects appear thus & thus situated one to another. The answer would be because they are really so situated among themselves & make their coloured pictures [images[a]] in ye Retina so situated one to another as they are & those pictures [images[a]] transmit motional pictures [images[b]] **into ye sensorium** in ye same situation & by the situation of those motional pictures one to another **the soul judges of ye situation of things without**. In like manner when we look with two eyes distorted so as to see ye same object double if it be asked why those objects appear in this or that situation & distance one from another, the answer should be because through ye two eyes are transmitted **into ye sensorium** two motional pictures by whose situation & distance then from one another **the soule judges she sees two things so situate & distant**. And if this be true then the reason why when the distortion ceases & ye eyes return to their natural posture the doubled object grows a single one is that the two motional pictures **in ye sensorium** come together & become coincident.[43]

In this passage we learn that in the *sensorium* there is a soul that 'judges of the situation of external things without', that is to say, 'she' (note the pronoun) apprehends that such and such is the case concerning their position and distance in external space. But is this soul another name for Newton's moveable principle, *anima,* or is it his other, immovable principle, *animus*? He does not say.

Nevertheless, the paragraph is of interest, because of Newton's attempt to address physiologically the problem of visual localisation, namely, how a spectator, using one or two eyes, discriminates the relative positions of visual stimuli (images[b]). His concern with this problem appears much earlier in his unpublished writings.[44] And in 1677 he considered physically, not physiologically, the problem of auditory localisation as a consequence of having been asked to comment on a version of the coincidence theory of harmony in *A Philosophical Essay of Musick*, published anonymously but written by Francis North.[45] The latter had supposed that even though two strings of a musical instrument may not begin vibrating at the same instant, resonance will work the strings into synchronism after the second or third pulse and, consequently, the listener will be affected equally by the two sound sources.

To criticise this supposition, Newton used the idea of intersecting waves as the appropriate analogue for sound propagation to conclude that

---

[41] See ibid., p. 382, where he compares the bent optic nerves to branches of 'Codlings' (trees with apples called codlings) twisted to form hedges or branches of fruit trees espaliered ('nailed up) to a wall'

[42] Ibid., p. 384.

[43] Newton, *The Correspondence*, vol. 2, Letter 261, pp. 383–4.

[44] See Newton, 'Of Colours', pp. 485–7.

[45] For the letter in which this consideration appears, see Kassler, *The Beginnings of the Modern Philosophy of Music*, pp. 175–8.

> ...the pulses of two sounds may at the same time strike one ear both together and another ear alternately and yet do exhibit the same concord to all ears, [so that] it follows 1st that concords arise not from the coincidence of pulses at the ear nor have any dependance [*sic*] on such coincidences, and 2dly that unisons are rather a harmony of two like tones then a single tone made more loud and full by the addition as the author would have it....[46]

Note, therefore, that in his letter to Briggs, he observes in passing that 'the situation of sounds depend not on their tone', that is, pitch.[47]

In Newton's day, the problems of auditory and visual localisation were intractable, and solutions were not provided until long afterwards. As for the problem of singleness of vision, any attempt at a solution would not bear fruit until sometime after the hypothesis on which it was based was brought into disrepute.[48] Indeed, Newton himself admits he is unable to explain how the 'coincidence' of the two images occurs, which, of course, means also that he had yet to arrive at his 1704 physiological explanation from the anatomical structure of the optic nerves and the continuity of their fibres.[49]

Nevertheless, in considering this problem, he offers some further criticisms of Briggs's 1682 tract. On the one hand, he asks whether that physician would have 'ye succus nutricius run along the marrow of the nerves' and cautions that, if his answer is affirmative, then 'tis an opinion not yet proved & so not fit to ground an argument on'. This statement suggests that by 1685 Newton had discarded Willis's nutritious juice (*succus nitritis*).[50] On the other hand, he points out that

> If you say yt in ye Camælion and Fishes ye nerves only touch one another without mixture & sometimes do not so much as touch; 'Tis true, but makes altogether against you. Fishes looke one way with one eye[,] ye other way with ye other: the Chamælion looks up with one eye, down with tother.... And in these Animals which do not look ye same way with both eyes what wonder if ye nerves do not joyne? To make them joyn would have been to no purpose & *nature does nothing in vain*.[51] But then whilst in these animals where tis not necessary they are not joyned, in all others wch look ye same way with both eyes, so far as I can yet learn, they are joyned. Consider therefore for what reason they are joyned in ye one & not in the other. *For God in ye frame of animals has done nothing without reason*.[52]

Here, then, we have ample evidence that the theory in Briggs's 1682 tract had stimulated Newton to reflect on, as well as to consider the comparative anatomy involved in the problem of singleness of vision.[53]

---

[46] Ibid., p. 178; for a comment on this part of Newton's letter, see pp. 98–100, 121, and for one response to his letter, see pp. 102–3.

[47] Newton, *The Correspondence*, vol. 2, Letter 261, p. 383.

[48] For the hypothesis, see *supra* Pt.II.2.1, p. 39.

[49] See *supra* Pt.I.12, Text VII (Qu. 15).

[50] See *supra* Pt.I.12, Text II, and *supra* Pt.II.2.1, p. 43. Note that after 1664 this hypothesis was widely adopted and sometimes modified.

[51] The source of this italicised phrase may have been an ancient anatomical maxim; see *supra* Pr.I.1.2, Texts VIII, p. 20 n.103 and XI, p. 25 n.133. Note that Newton provides an important gloss on this phrase in the sentence italicised at the conclusion of the quotation.

[52] Newton, *The Correspondence*, vol. 2, Letter 261, p. 384 (italics mine).

[53] See *supra* Pt.I.1.2, Text VII (Qu. 40); see also *supra* Pt.II.2.1, p. 40 and n.43.

After a hiatus of some months, Briggs published his second, 1683 tract in which he sought to defend his new theory. Consequently, he began with a summary of what he had maintained in his 1682 tract; and he followed this summary with answers to seven objections that had been sent to him 'by Mr. *Newton* our worthy Prof. of *Mathematicks* at *Cambridge*', as well as by some unnamed friends.[54] These answers contain a number of statements that deserve notice, the first of which is that 'in our view of an object the true situation of the respective parts is not distinguish't so much by the means of that end of the ray that *terminates in the Eye* as of t'other end that *touches the* Object, from whence the vibration or *protrusion* comes.[55]

Does this statement suggest that there is *no* transport of matter from the object to the eye? That is, does it suggest that what is transmitted is a physical state?[56] If so, it would indicate a change from his 1676 treatise, in which Briggs conceived light rays as an emission of unobservable particulate substances that enter the eye and strike the coat of the retina. However, evidence for such a change of view is somewhat ambiguous in his later 1683 tract. For, on the one hand, Briggs used the phrase, 'the finer stroaks of the *etherial or lucid matter*' in the retina, which suggests the transport of matter.[57] And on the other hand, evidence that implies the transmission of a physical state follows from one of the main objections to his new theory—'That it seem'd difficult to conceive how those *soft Medullary Fibres* of the [optic] Nerve could have such a *tension*'.

In answering this objection, Briggs compared the fibres under the retinal coat to the tension in the web of a spider,

> ...whose *Mucous substance* and *Expansion* very well answers to that of the *Retina*, whilst in its due position or Expansion in the Eye; and as the least *breath of Wind* moves the one [the web], so the least gale of the *Etherial or lucid matter* causes a *vibration* in the other.[58]

And at the conclusion of his tract, he returned to this analogy when answering a final objection that the retina's substance is not suited to vision.[59] His answer to this objection starts from the premiss that '*Vision* can be no way better performed then

[54] Briggs, 'Continuation', p. 175.

[55] Ibid., p. 177.

[56] Note that this would imply a continuum, not an emission hypothesis; see Shapiro, 'Kinematic Optics', p. 136 n.5.

[57] Briggs, 'Continuation', p. 181.

[58] Ibid., pp. 175–6.

[59] The objector seems to have been Edme Mariotte, who in 1668 described the blind spot and localised it at the point of emergence of the optic nerve from the retina; see Boring, *Sensation and Perception*, pp. 100 and 121. It is puzzling, therefore, that he also argued against the retina and for the choroid ('chorœides') as the peripheral end-organ of vision; see 'The Answer of Monsieur Mariotte to Monsieur Pecquet'. Today, the choroid, with the iris and ciliary body, is considered as part of the uvea, the pigmented, vascular layer of the eye. In addition to the iris, Briggs, *Ophthalamographia* [1676], described the other two bodies, the first of which, p. 17, is referred to by its Greek name, the last of which, p. 18, is referred to as 'processus ciliares'. The 'processes' of the latter, the *ligamentum ciliare*, are given a short, separate description, pp. 22–3, in which Briggs stated that they are arranged like the teeth of a comb. It is possible, therefore, that this description could have influenced Newton's choice of a comb to make the experiments described *supra* Pt.I.1.2, Text V.

by the *Fibres of the Retina*'. But before resuming his spider-web analogy, he offered three supporting reasons for his premiss, including the following:

> First ... the *Retina* is no more transparent if so much as the *oil'd paper* in the *lantern*, which yet serves well enough to intercept the *turning images* of it. Secondly, ... being of a *whitish colour*, and resembling thereby the *white Paper* in the dark house, it is fitter to receive the images of *colour'd objects*.... Thirdly, It being the more inward or *medullary expansion* of the *Optic Nerve* it can more Immediatly [*sic*] transmit any motions to **the *Meditullium*[60] of the brain (or the *common sensory*)**....[61]

Note that the second point enumerated in the passage quoted above contains the only allusion in Briggs's writings to a *camera obscura*. More important, however, is the third point, in which we learn that the substance of the retina is 'whitish'; that the substance of its 'inward' expansion is white (medullary); and that the '*common sensory*' is located in the center or middle of the under surface of the brain, that is to say, in its white (medullary) center. Is there, then, some kind of analogy between the retina and the common sensory? The answer seems to be yes, for in returning to his spider-web analogy, Briggs stated that

> ...the *Retina* has as much [tension] as those of a *Spider's Web*, and this is sufficient, nay more suitable to the finer stroaks of the *etherial or lucid matter* and the *nice actings* of this sense, which is not required in any other: and therefore the same objections may lye against the *Constitution of the brain* it self, which consists of soft *medullary Fibres* that are however fit enough to receive or propagate any motion, and whilst they are fill'd with *Animal spirits* may be allow'd to have the like tenseness or resistance that a *lock of Wool* has, or a Spider's Web. And if I may be admitted to carry on the similitude; As that *little Animal* in the *Centre* of its soft *circumtended Fibres* is sensible of the least gale of Wind ... by the *delicate expansion* of its Fibres: So may **the Soul** much more (**in the *common sensory***) being *surrounded by Fibrillæ*[62] of *expanded Nerves* and of a *finer make*, **apprehend** from what Quarter the several motions come from abroad, **and more *minutely perceive*** *the difference* of 'em in respect of the *diverse Organs* of sense and the different fineness or *tension* of those Nerves that belong to the same.[63]

Just as in Newton's September 1682 letter, in which the soul ('she') is in the *sensorium*, so too in the passage quoted above, the soul is 'in the *common sensory*'. But from Briggs' spider-web analogy, we may infer that the soul *feels* the situation and distance of external objects from stimuli transmitted to the common sensory from the peripheral end-organs for vision, hearing and other sensory modalities. If this inference is correct, then the soul is a sensitive substance *in* the common sensory and, hence, is analogous to the light-sensitive substance *in* the retina, for in both the retina and the common sensory, there is an expansion of fibres that contain animal spirits.

It should be noted, therefore, that in his 1676 treatise, Briggs clearly stated that 'the composite *spiritus animales*' constitutes 'the *entire corporeal anima*',[64] the

---

[60] OED obs.: The center of anything; here, part of the brain (encephalon) called 'the cerebrum'.

[61] Briggs, 'Continuation', p. 175.

[62] I.e., the off-shoots or subdivisions of nerve fibres.

[63] Briggs, 'Continuation', pp. 181–2.

[64] Briggs, *Opththalamo-graphia* [1990], p. 34.

'nature and movement' of which, 'like the *spiritus chymici*', may vary as to quality and operation.[65] Consequently, it is this unobservable corporeal substance that constitutes Briggs's principle of sense and motion (*anima*). But he also believed that 'some *clear principle* [*animus*] seems to exist in us, which, as a participant of the divine spirit, exercises a modifying power in our perception'.[66] Accordingly, he posited two souls, one, corporeal, the other, incorporeal; and in doing so, he revealed a covert debt to Willis, as will soon be evident. But Briggs's debt to Willis is also evident in a number of other ways.[67]

For example, he retained the latter's idea of a nutritious juice (*succus nitritus*) running along the marrow of the nerves, even after Newton's 1682 warning that this juice was 'not proved'.[68] He also relied on Willis's ordering of the cranial nerve-pairs, since he mentioned the first (olfactory), second (optic), third (oculomotor), fourth (trochlear), fifth (trigeminal) including one of its branches (opthalamic), sixth (unnamed but now called 'abducent'); and he also referred to the spinal and 'intercostal' nerves, the latter being Willis's name for the sympathetic trunk.[69]

More important for our inquiry, however, is Briggs's description of the origin of the optic nerves in the brain quoted previously. For although his description lacks precision, he clearly was indebted to Willis's more precise description, that 'the Optick or seeing Nerves ... arise from [the optic chambers] *behind* the chamfered bodies out of the second section [or oblong marrow] of the medullar shanks',[70] the 'chamfered bodies' being one of the names for the corpora striata given by Willis or by his translator, Samuel Pordage. And in an illustration (see Fig. III.2), showing the oblong marrow removed from a sheep's brain, the fore-mentioned bodies are positioned above the letter K, where the two optic chambers (HH) are represented as lying *under* the corpora striata (A B), the two bodies in the 'Meditullium' of the brain that Briggs identified as the common sensory.

The striated body on the left (A) has been cut through the middle to expose its 'chamferings' or striations, whereas the striated body on the right (B) is uncut and, hence, shows the 'brim' of the corpus callosum (C C) '*cleaving to the chamfered body*'.[71] In anatomical description, therefore, words such as 'behind', 'in front',

---

[65] Ibid., p. 45

[66] Ibid., p. 34.

[67] See ibid., pp. 6, 28–30, 41 (citations to Willis, *Cerebri anatome* [1664], Ch. 21 and 22, on the cranial nerves); and Briggs, 'A New Theory', p. 176 (citation to Willis, *De anima brutorum* [1670], Ch. 15, on vision). See also *supra* Pt.II.2.1, p. 46.

[68] See Briggs, 'A Continuation', p. 174.

[69] See Willis, *The Anatomy of the Brain* [1664 tr. 1681], pp. 137–41, pp. 141–48, pp. 157–73, and pp. 173–6. See Briggs, *Opththalamo-graphia* [1990], first pair, p. 30; second pair, *passim*; third pair, pp. 2–4, 7, 10, 26–30, 52; fourth pair, pp. 6–7, 26, 32, 51, 52; fifth pair, pp. 33–4 (implied), 40–1, 52; sixth pair, pp. 7, 52; spinal nerves (Willis identified 31 pairs), p. 33; and intercostal nerves, p. 41.

[70] Ibid., p. 139 (italics mine). The word 'shanks' here refers to Willis's 'medullar trunk', which includes both the oblong and spinal marrow; see pp. 82–3 below.

[71] Ibid., p. 105 (explanatory table to Fig. 8); see also pp. 61–2, where is indicated that the same method of dissection may be used for the human brain. But Willis, *Two Discourses* [1672 tr. 1683], Fig. 8 facing the explanatory table, p. 44, used a different method to expose some of the same parts in his 'new Anatomy of the Human Brain'.

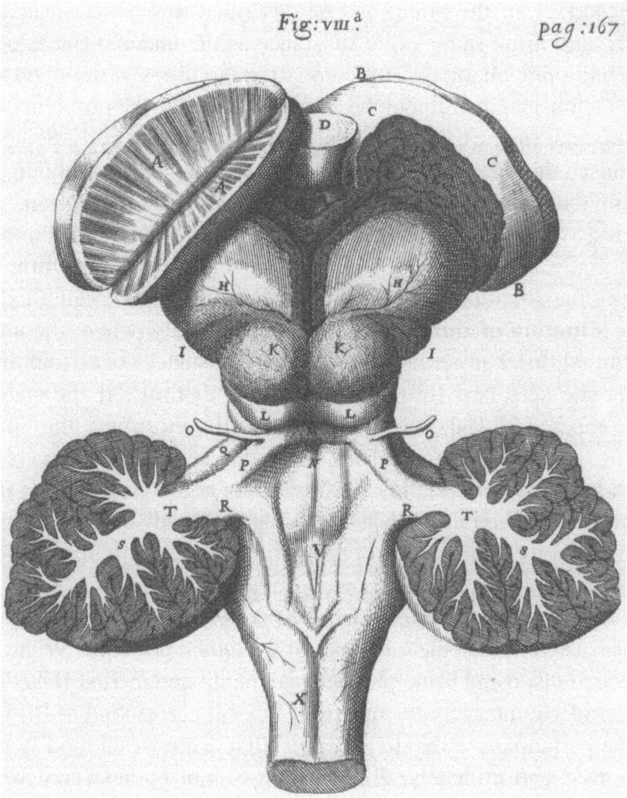

Fig. III.2 Oblong marrow (medulla oblongata) removed from a sheep's brain *from* Willis, *The Anatomy of the Brain* (1664 tr. 1681), Fig. 8

(above K K)
H H Pair of optic chambers
A B Pair of corpora striata, in which (A) is cut through the middle and (B) is uncut
C C That part of the coat of the corpus callosum cleaving to (B)

'above', 'under' depend on the method of dissection used to expose parts in the white medullary under surface of the brain. Once this is recognised, the problem identified earlier with the two unnamed bodies in Briggs's illustration is no longer a problem, for they are indeed the corpora striata (see Fig. III.1, letter *h*, above).

3.2 Thomas Willis on the Nervous System

In the previous section, new details came to light concerning what Briggs called 'the Meditullium of the brain (or the common sensory)'. For in this equation he hinted that the location of the common sensory is somewhere in the center ('Meditullium') of the under surface of the brain; and he described this under surface as fibrous ('medullary') and of the same white substance as the under surfaces of the retina, optic nerves and optic chambers. By comparing the fibres in the optic nerves with taut strings of a musical instrument, he attempted to solve the problem of singleness of vision (binocular fusion). But in a letter to Briggs, written on 12 September 1682, Newton criticised this analogy as an inadequate solution to the problem, at the same time admitting that he himself was unable to offer a different solution.

Nevertheless, that letter contained two statements of particular importance for this inquiry: firstly, when images[a] are transmitted '**into ye sensorium** in ye same situation & by the situation of those motional pictures one to another[,] **the soul judges of ye situation of things without'**, and secondly, when 'the motional pictures' transmitted differ in terms of 'situation & distance' one from another, '**the soule judges she sees two things so situate & distant'**. If the statements just quoted are compared with the much later Text XII concerning 'the Soul of Man', then the term 'soul' in Newton's 1682 letter denotes an unobservable corporeal substance, the property of which is sensitivity. As noted previously,[72] in Text II of 1675/6 Newton conceived this substance as part of the distributed ætherial animal spirits in the nervous system and, hence, as that part of his disposing principle, *anima,* concerned with performing sensory-motor functions.

Evidence has also been adduced that for performing these functions, Willis, as well as Briggs also relied on the same part of the *anima* principle. What, then, about Newton's second disposing principle, *animus*, mentioned in Text II as the unknown efficient cause of voluntary motor function? Does it correspond to Briggs's second principle, 'the participant in the divine spirit' that 'exercises a modifying power in our perceptions'? Unfortunately, this question cannot be answered with absolute certainty from Newton's texts. Fortunately, however, the work of Willis, which is the subject of this section, will provide considerable insight into how the question might be answered. In addition, from his detailed description of the common sensory and its specific place in the anatomical space of the brain, we shall learn that it is the seat of sensory as well as motor function. And we shall also discover that it is isomorphic or structurally similar to the retina in the interior space of the eye.[73]

In terms of his own generation, Willis was regarded as the most prominent physician after William Harvey, whereas history remembers him for putting the mechanisms of the brain and nervous system on their modern footing, insofar as could then be done.[74] It is known with certainty that Newton owned one book by Willis,[75] and there is a high probability that he read a number of his other books. But if he read those books carefully, he would have recognised that Willis had a project that

---

[72] See *supra* Pt.II.2.1, p. 47.

[73] See pp. 84–6 below.

[74] This section is partly based on Kassler, *Music, Science, Philosophy*, pp. 124–65, and sources cited there.

[75] See *supra* Pt.I.1.2, comment to Text II, comment to Text IV and comment to Text X; see also *supra* Pt.II.2.1, p. 45.

resonated with his own plans to banish Cartesian notions from natural philosophy.[76] For although, like Newton, Willis had benefitted from his study of the writings of René Descartes, yet he rejected the Cartesian mechanical philosophy, interpreting it as pure mechanism and as mind-body dualism carried too far, thereby posing the ultimate theological problem of God's power and presence in his creation.

In order to harmonise natural philosophy with his Anglican faith, Willis sought to reanimate the human machine by positing two gendered souls[77]: one, a disposing principle of animal life, sensation and motion (*anima*) that is 'traduced' by, and has her gifts from the male parent; the other, a disposing principle of intellectual life (*animus*) that has his gifts from the deity.[78] By conceiving the former soul as an unobservable corporeal substance that is non-rational, mutable and mortal and the latter soul as an unobservable incorporeal substance that is rational, immutable and immortal, Willis was able to speculate on highly controversial, pathological aspects of psychosomatic life previously regarded as the domain of the theologian.[79] In these speculations Willis followed a broadly Platonist tradition, probably derived from Greek naturalists such as Galen,[80] for as a medical graduate of Oxford University, he apparently had no direct connection with the circle of philosophical Platonists at Cambridge University.[81] Unlike that circle, however, he recognised that speculations required an empirical foundation from investigations into the anatomy

---

[76] See *supra* Introduction p. viii. For extensive evidence of Newton's plan, see Guicciardini, *Isaac Newton*; for a near contemporary who grasped the nature of that plan, see Kassler, *Seeking Truth*, p. 54, *et passim*; and for some of Newton's difficulties in accomplishing it, see *infra* Conclusion, pp. 134–5.

[77] Although Willis developed the dual-soul concept for his own purposes, yet his older contemporary, Pierre Gassendi, also had recourse to a similar concept to serve his own purposes. For since the latter aimed to reconstruct the materialist teaching of the Epicurean atomists, as a divine he had to avoid reducing the principle of *animus* to material soul atoms. And it was this requirement that led him to propound a double composition of the soul. See Jones, *The Epicurean Tradition*, pp. 55–61; see also Osler, 'Baptizing Epicurean Atomism'. For some ancient antecedents for the dual-soul concept, see Kassler, *Music, Science, Philosophy*, pp. 64 n.10, 65–6. 147–8; see also Lee (tr.), *Plato Republic*, Pt. V, Bk. IV, pp. 206–22, and Lawson-Tancred (tr.), *Aristotle De anima*, II.ii–iii, pp. 158–64.

[78] Willis, *Two Discourses* [1672 tr. 1683], p. 42. Of the two discourses, (i.e., medical exercises or practices), one is physiological, intended to show the nature, disposing powers, affections and parts, vital and sensitive, of the corporeal soul, whereas the other is pathological, intended to unfold the diseases which affect the 'primary seat' in the brain of the sensitive part of this soul. Despite his focus on *anima*, there are scattered references throughout the book to *animus*.

[79] For examples, see Willis, *An Essay of the Pathology of the Brain* [1667 tr. 1683], pp. 11 (the 'holy' disease, epilepsy), 43–4 (the 'universal convulsions from witchcraft'). See also Kassler, *Music, Science, Philosophy*, pp. 127, 137–8, 151 and n.73, 152–3, 158–60.

[80] See ibid., pp. 127, 147–8, 151 and n.73, 152–3. For the pre-Galenic medico-biological context of the Hippocratic corpus and for Plato's theory in the *Timaeus* and Aristotle's development of that theory, see Tracy, *Physiological Theory and the Doctrine of the Mean*.

[81] Like the Platonists at Cambridge, Willis embraced some form of atomism, probably having learned about it from Gassendi, whom he cites in *De anima brutorum* [1672], and perhaps also from one of Willis's patients, Anne Conway, a pupil of the Cambridge Platonist, Henry More; see Dewhurst, 'Some Letters', pp. 74–5.

of the nervous system,[82] in the course of which Willis brought about a mechanico-chemical synthesis.[83]

The nature of this synthesis will be clear from a brief outline of his approach to four inter-related areas of investigation, the first two of which modify the Galenic anatomical tradition. First, there is a general anatomy that identifies the organs and their structural connections, the organs being conceived as containers with pores.[84] Second, there is a functional anatomy that identifies the offices and uses of these organs, conceived as different kinds of mechanisms. Third, there is a chemical anatomy that analyses the blood into five chemical 'principles'—spirit, sulphur, salt, water and earth.[85] And fourth, there is a physiology of the particulate animal fluids, which, in the containers, undergo chemical processes used in distillation for separating substances according to their volatility.

For Willis, blood, the principal fluid in the body, is the matter to be distilled. During its circulation, chemical transformations result from a micro-mechanical interaction of its unobservable corporeal particles, called 'fermentation', which, as indicated in Part II, is a boiling motion of the particles that is accompanied by heat. But he supposed that the interacting particles themselves correspond to the five previously-mentioned chemical principles, of which the most important is spirit, for this is generated in the heart as vital spirit and reaches perfection in the cortex of the brain as animal spirit. Willis, therefore, based his physiology on two main concepts, fermentation and spirit.[86]

All four investigative approaches assisted Willis in developing his substance conception of the two souls and their respective knowing powers,[87] the latter of which he

[82] Willis's speculations were based not only on anatomical dissections but also on 'farmyard empiricism', chemical studies, experiments and clinical experience, for details of which, see Dewhurst, *Willis's Oxford Casebook*. Unfortunately, there is a tendency among some commentators to criticise Willis for his speculations; but they do so without taking into account not only the problem of a developing field that requires new terms, but also the nature of anatomical reasoning. For the latter, see Stevenson, 'Anatomical Reasoning in Physiological Thought'.

[83] Consequently, Willis cannot be classed with either the iatromechanists or the iatrochemists.

[84] Willis used various terms for the containers, including chambers, cisterns, cloisters, pipes, repositories, reservoirs, sinks, storehouses, tubules.

[85] This division modifies the three-element theory of Paracelsus, in which mercury is renamed as spirit; see Bastholm, *The History of Muscle Physiology*, p. 203. For his description of the five principles, see Ch. 2 in Willis, *Diatribæ duæ medico-philosophicæ* [1659]. For his belief that they are directly implicated in disease; see Davis, *Circulation Physiology and Medical Chemistry*, p. 154 *passim*.

[86] According to Willis, *The Anatomy of the Brain* [1664 tr. 1681], p. 98, 'all the Pores and passages in all the parts of a living Creature are dilated by spirit and heat', i.e., the motion of fermentation Cf. Newton *supra* Pt.II.2.2, pp. 47–8 and Briggs p. 62 above.

[87] According to Wilkie, 'Body and Soul in Aristotelian Tradition', p. 20, for Aristotle and his successors, the problem of the relation between soul and body was 'never felt to be particularly troublesome', whereas the 'really difficult problem for them was the relation between the *imagination* and the *intellect*.'

abbreviated as 'Intellect' and 'Phantasie'.[88] According to his doctrine (and anticipating somewhat), intellect, a rational power of the superior incorporeal soul (*animus*), is conceived as an immovable, unextended, indivisible 'pure' spirit.[89] Located only in the brain, this power mediates cognitive functions of perceiving and thinking, as well as rational will, a faculty directed to rational action. Accordingly, there is an identity between intellect and rational will.[90] Phantasie, a non-rational power of the inferior corporeal soul (*anima*), is conceived as a movable, extended, divisible body of spirits. Located partly in the blood and partly in the brain, this power mediates vital and sensitive functions due either to passive-sensory or active-motor 'impulses'. The passive-sensory are twofold: vital impulses due to intrinsic stimuli within the body and natural impulses due to external stimuli. The active-motor are 'animal' impulses due to 'sensitive Appetite', a faculty directed by phantasie to conscious and intentional action. Accordingly, there is an identity between phantasie and sensitive appetite. But there is no identity between intellect and phantasie.[91]

From these preliminary remarks, it will be obvious that Willis's natural philosophy is rather complex (and I have only scratched the surface).[92] But the distinctions made above are intended to assist the reader in what is to follow, beginning with Willis's functional anatomy of the eye,[93] which he claimed was taken 'not from the Writings of others, but by our own ocular Inspection, and observation of the Eye and its parts'. In the just-cited passage the term 'parts' refers to Willis's division of the eye into outer or 'extrinsic', inner or 'intrinsic' and interior or 'inmost'. The extrinsic parts include the 'round Bone', the eyelids with their hairs, the eyebrows, the '*Glandula* or Kirnels, with the Vessels, and Excretory passages', whereas its

[88] Note that this part of Willis's dual soul concept also has ancient antecedents that include the tradition of interpreting allegorically the biblical 'Fall' of man in Genesis. For example, the Greek-speaking Jewish-Platonist philosopher, Philo of Alexandria, repeatedly suggested that the first man, Adam, symbolises intellect (*nous*) and the first woman, Eve, sensitivity (*aisthēsis*); see Stead, *Doctrine and Philosophy in Early Christianity,* IX, p. 230. According to Lawson-Tancred (tr.), *Aristotle De Anima*, p. 118, the Greek word, *aisthēsis*, is a term of wide meaning, encompassing the sense of three modern terms: sensation, perception and consciousness.

[89] Willis, *Two Discourses* [1672 tr. 1683], p. 41. The doctrine that incorporeal substance is indivisible has a long history, starting with the *Parmenides* of Plato, taken up, e.g., by Philo of Alexandria and 'vividly characterised' by Origen; see Stead *Doctrine and Philosophy in Early Christianity*, V, p. 44.

[90] A doctrine of the identity of intellect and rational will was commonly held by seventeenth-century Platonists in opposition to the received Thomistic doctrine of the real distinction of these powers. For example, in discoursing on the mind as the 'seat or subject of faith', the Rev. John Tillotson claimed to make no 'real distinction of Faculties'. But if some person would distinguish them, 'the proper seat of this persuasion [i.e., faith] is the Understanding [i.e., intellect]; the immediate effect of it is upon the Will; by which it works upon the affections and the life'; see Dockrill, 'Enthusiasm in Seventeenth Century England', p. 151.

[91] Willis *Two Discourses* [1672 tr. 1683], pp. 41–2.

[92] This complexity has not been grasped by Wallace, 'The Vibrating Nerve Impulse', who therefore has missed the multiple nature of Willis's dualism, as well as its ramifications for his anatomy and physiology.

[93] See ibid., Ch. XV ('Of the Sight'), pp. 75–86. All quotations concerning Willis's ocular anatomy are taken from this chapter.

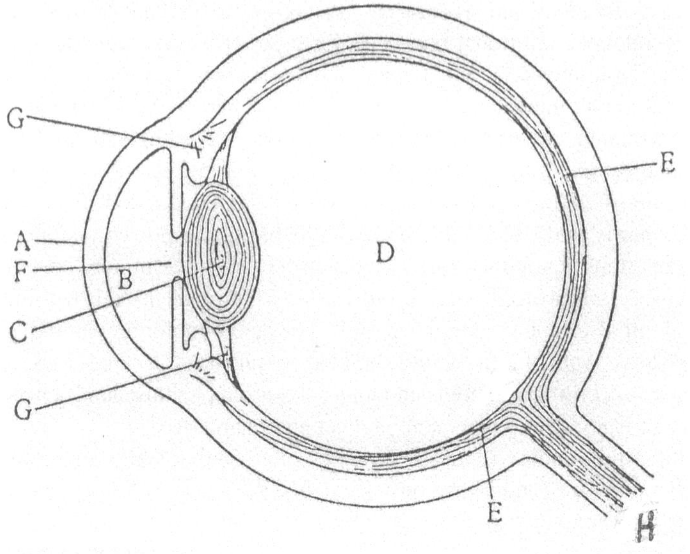

Fig. III.3 Innermost part of the eye *after* the description in Willis, *Two Discourses* (1672 tr. 1683)

A	Coat of the cornea
B	Space filled with aqueous humour
C	Crystalline humour
D	Space filled with vitreous humour
EE	Coat of the retina
F	Pupil
G	Ciliary processes
H	Optic nerve cut off

intrinsic parts include the 'Globe' or eyeball itself, as well as parts disposed 'about its Compass'—the muscles and vessels, with the fat lying between. For the purposes of our inquiry, the most important division is the eye's 'Penetralia or inmost parts', namely, its five coats and three humours (see Fig. III.3).

Of the five coats, three are 'greater' (so-called because they are 'stretched' throughout most of the interior space of the eye) and two are 'lesser', that is, occupy less space. The lesser coats are (A) the cornea ('horny' coat) in front of (F) the pupil, and the 'inmost' coat (E), the retina, whereas the greater coats[94] serve as containers for three humours—(D) vitreous ('glassy'), (C) crystalline and (B) aqueous ('watry'). The iris surrounding the pupil is made up of (G) '*Ciliare* Processes',[95] a 'Fringe' of

[94] These are the sclerotic or outermost coat, the albuginia and the choroid, the last of which is rendered by Willis's English translator, Samuel Pordage, as 'chorocœides' or 'crocoideos'.

[95] According to Willis, these ciliary fibres, which stick to the cornea, are contiguous not only to the crystalline humour and its little membrane called 'the Cobweb', but also to the retina; and they have two uses: (1) to expand and contract the pupil, and (2) to 'thrust forward, or draw backward the Chrystalline humour, and bend it hither and thither into the view of the [external] Objects'. In his description of the second use, Willis was on the right tract, for his ciliary processes are now called 'ciliary muscles', and their function is to adjust the power of the crystalline lens.

variegated fibres carried from the pupil 'like rays from a luminous body'. And exiting behind the retina is (H) the optic nerve (cut off).

From the pupil, there are various other apertures (holes') that serve for the transmission of stimuli from an external object to the retina. However, for image[a] formation Willis required that the coat of the retina be struck by real but unobservable *corpuscula* or little atomic bodies,[96] conceived as hard rays of light, the chemical property of which he described as 'fiery', that is to say, nitro-sulphuric. For

> ...the structure of every Sensory [i.e., peripheral end-organ] is so made, according to its Pores and Passages, that Particles only proportionate to them may be admitted in: wherefore as Light, and the Images of things, pass thorow Glass, and clear Bodies; not dark Bodies; so the same are received only by the Eyes, and not by the other Sensories.[97]

As the most moveable, as well as 'Active' particles, the light rays, which are transmitted to the coat of the retina instantaneously (in a 'twink'), pass through glass and other solid 'Diaphanous Bodies' but not opaque ones. But what happens once the rays enter the interior space of the eye?

Willis's answer was not original with him, for he acknowledged that the eye's mechanism was commonly, 'and not improperly', declared

> ...by the example of a Burning-Glass,[98] which like a little Window is fixed before an hole made in the Wall of a shut up and dark Chamber; Because, from the [external] Bodies everywhere brought before that hole, the Rays of Light being reflected, meet together in the Glass, and in that passage cutting one another, spread themselves at last within the Chamber, and so upon a white Wall within, represent a Landschap [i.e., landscape] of the whole visible Hemisphere.[99]

Willis, therefore, conceived the interior space of the eye as an optical instrument. According to this model, the 'hole'—eye-glass or ocular— is the pupil at the centre of the iris, which controls the amount of light entering the optic tube of the dark chamber. The chamber's four walls represent four coats, whereas the white wall represents the fifth coat, the retina. Finally, three convex lenses or 'Dioptrick Glasses' represent the transparent humours: the aqueous humour; the crystalline humour (the burning lens); and the vitreous humour (a 'fused' lens[100]).

Once the rays are admitted through the ocular of the optic tube, these lenses dispose the rays according to 'the best Dioptrick Rules' and, hence, 'aptly break

---

[96] This is Willis's requirement for all sensory modalities; but he also required that the chemical properties of the atomic bodies be different for each sensory modality, which, of course, have different functions.

[97] Willis, *Two Discourses* [1672 tr. 1683], Ch. X (Of the Sense in General), p. 57. Unlike Newton, Willis distinguished between the terms 'sensory' as a peripheral end-organ and 'common sensory' as a central end-organ.

[98] I.e., a convex lens; see Newton, *Opticks (1)*, p. 6.

[99] Willis, *Two Discourses* [1672 tr. 1683], pp. 77–8 *et passim*. In 1658 the Norfolk physician, Thomas Browne, modelled the functional anatomy of the eye on a *camera obscura*; and in 1659 Willis first mentioned the *camera obscura* in connection with sight; see Willis, *A Medical Philosophical Discourse of Fermentation* [1659 tr. 1681], p. 39 ('close chamber').

[100] Scheiner may have been Willis's source for a 'fused' lens; see Crombie, *Science, Optics and Music*, p. 232 n.100 and p. 233 (Fig. 36).

[refract], and gather together, the Beams', thus exhibiting the images[a] on the retinal coat. From there, the images[b] are 'delivered' to the optic nerves that exit behind the retina. Thus, 'the Eyes and Optick Nerves belonging to them, perform the whole Act of Seeing', for once 'the Images of all visible things are formally painted' on the retina, they are afterwards conveyed through the pores and passages of the optic nerves to the common sensory. But where is the location of this entity?

Answers to the question may be found in what is undoubtedly Willis's most important book, *Cerebri anatome*. For prior to its publication, and even for sometime afterwards, seventeenth-century philosophers and physicians adhered to the long tradition, mentioned briefly in Part II, that relegated psychic faculties to the ventricles in the anterior and posterior portion of the brain (encephalon), that is to say, three ventricles in the cerebrum and one ventricle, the fourth, in the cerebellum.[101] Nevertheless, a number of modifications were made to that traditional theory. Recall that Henry More had identified the fourth ventricle as the place of the centre of perception or 'Common *Sensorium*', as well as the seat of the soul and its psychic faculties (i.e., animal spirits).[102] But Descartes, who did not use the term, '*sensorium commune*', directed animal spirits from the ventricles into the pineal gland, because he believed that this gland was the seat of the faculties of common sense and imagination and, hence, the place in which 'the rational soul', when united to the body 'machine', will consider 'directly' ideas, forms or images.[103]

Willis, however, dismissed Descartes's opinion, as well as the 'ancient Opinion', which some in his day had 'endeavoured to bring into vogue'.[104] Instead, he followed 'Anatomists ... of a later Age' who treated the ventricles as sinks to the 'more inward chambers of the Brain'. But from his own dissections, he concluded that these sinks receive the 'excrementitious', 'oftentimes deadly Humors' and excrete them into the pituitary gland through a passage called 'the Tunnel'.[105] Two years later, Newton, without mentioning Willis's name, wrote that in the 'center' or 'mi[d] st' of the brain, where 'the substance here is most pure', there is 'that small passage twixt y[e] ventricles where all superfluous [serous] humors have the greatest advantages to slide away that they may not incumber y[t] p'cious organ'.[106] Given the fact that the ventricles are parts in the brain, Newton's last three words suggest that he is referring to a precious organ in the middle of the brain and not to the ventricles or to the brain itself.

---

[101] For Willis, the two main parts of the brain (encephalon) are the cerebrum and cerebellum; see further pp. 81–2 below.

[102] See *supra* Pt.II.2.2, p. 56.

[103] See Descartes, *Treatise on Man*, p. 86, who, earlier in *La Dioptrique*, had identified 'the seat of common sense' as 'a certain small gland which is found about the center of these concavities [the ventricles]'; see Descartes, *Discourse on Method*, p. 100. The modern term is 'pineal body', which is attached to the roof of the third ventricle.

[104] Willis, *The Anatomy of the Brain* [1664 tr. 1681], p. 97.

[105] Ibid., pp. 96–100, 104, 123, *et passim*.

[106] I.e., precious; see Newton, 'Of Colours', p. 487, whose description (above) suggests that he actually read the first edition of Willis's *Cerebri anatome* [1664], a copy of which was in Barrow's library; see *supra* Pt.II.2.1, p. 39 n.39.

However, the important point here is that Willis, by providing a new use for the ventricles, was free to localise psychic faculties in the brain's two surfaces: memory in the upper surface of grey matter (the cortex) and common sense and imagination in the under surface of white matter. Since for him, function is inferred from anatomical fact, I shall provide a brief account of some aspects of his general anatomy, which is based on the model of a *hydro-pneumatic* organ.[107] But this model should not be confused with the structural analogue used by Descartes to explain how the brain receives impressions of external objects through the mediation of the nerves. For, on the one hand, his analogue was a carillon controlled by the keyboard of a *pneumatic* organ; and, on the other hand, his explanation of function was based on the premiss that the concealed motive power of animals is controlled by laws that govern automatic machines and, hence, is *involuntary*.[108]

Willis's hydro-pneumatic model is completely different, for it has four main parts: (1) air feeders and reservoir, complete with operating mechanisms; (2) windchest, combining soundboard and pipe rack with pipes; (3) a barrel pricked with automatic tunes; and (4), a key-frame and action parts. Of the four parts, the first constitutes the operating system with its hydraulic (water) and pneumatic (blowing) mechanisms, whereas the remaining three parts constitute the control system with its key-frame (representing the cerebral cortex), barrel (representing the cerebellar cortex) and windchest mechanisms (to be discussed below).

The mechanisms of the operating (cardiovascular) system and the control (nervous) system are used, respectively, for vascular work and nervous work. But the source of this work is the blood, an hydro-pneumatic principle conceived as 'chemical work', the distillatory processes of which are completed when the blood is regenerated in purified form as vital spirits in the heart and then as animal spirits in the brain,[109] the latter of which 'inspire' the windchest, as well as the nerve pipes and nerve fibres that innervate the muscles. Accordingly, nervous work is performed by the products of distillation that Willis called 'animal spirits' and, metaphorically, 'wind'.

Since these spirits are implanted in, circulate or flow through the containers of the nervous system, I shall focus on the control system,[110] which incorporates three principal mechanisms: the cerebrum, the cerebellum and their common appendix called the 'medullar Trunk' (the windchest). But before describing the connections

---

[107] For a full treatment of Willis's model, see Kassler, *Music, Science, Philosophy*, pp. 125–64.

[108] See Kassler, 'Man—A Musical Instrument', pp. 62–6; Kassler, *Inner Music*, pp. 43–8; and Kassler, *The Beginnings of the Modern Philosophy*, pp. 9–10. Descartes and, before him, Fernel were forerunners of the concept of reflex action, but Willis's contribution was more significant; see Brazier, 'The Historical Development of Neurophysiology, p. 31.

[109] Harvey, *The Anatomical Exercises*, p. 157, had condemned spirits as a *deus ex machina* and the recourse to them as an asylum of the ignorant. Perhaps in answer to Harvey's ghost, Willis, *Two Discourses* [1672 tr. 1683], p. 23, wrote: 'I shall say nothing to those, who wholly deny these Spirits, for ... the exi[s]tences of which, is almost palpable, and may be proved demonstratively by the effects.' Note his indebtedness to the method of physiological explanation of Galenic tradition described *supra* Pt.II.2.2, p. 52.

[110] For details of the operating system, see Kassler, *Music, Science, Philosophy*, pp. 129–40.

between these mechanisms, it is important to note that two features are unique to Willis's control system. On the one hand, the cerebrum and cerebellum constitute a *dual* control system.[111] On the other hand, this system has its corollary in a *dual* memory system, one part of which, located in the cerebral cortex, controls voluntary ('spontaneous') function, the other part of which, located in the cerebellar cortex, controls involuntary ('instinctive') function.[112]

Because of Newton's focus on voluntary action in connection with animal motion, the cerebellum falls outside the remit of this inquiry,[113] even though the modern history of this mechanism begins with Willis.[114] For his general anatomy, therefore, I shall merely indicate that although the cerebellum is attached to the hinder part of the cerebrum, yet these two mechanisms have no immediate communication, one with another. But they do have mediate communication by means of the medullar trunk (see Fig. III.4).[115] This mechanism not only includes the oblong marrow (medulla oblongata) but also four 'orbicular protuberances',[116] the pons of Varioli and the complete spinal marrow (medulla spinalis).[117]

As the oblong marrow, the trunk originates in the cerebrum where (A) the corpus callosum ends and (B) the two convex 'prominences' of the corpora striata begin.[118] The latter form 'the tops' of the oblong marrow, which then extends downwards past (C) the two optic chambers (thalami optice). From there, it enters into (D) the four orbicular protuberances, (E) the pons, (F) the rest of the oblong marrow to (G) the spinal marrow, the last of which extends all the way to the os sacrum, the ridge on which the backbone rests (not represented in Fig. III.4).

---

[111] This feature is also unique to the musical instrument that is the probable source of Willis's model; see ibid., pp. 160–4.

[112] In the medical tradition that Willis inherited, a distinction was made between three classes of motor function: voluntary ('animal'), involuntary ('vital') and movements partly under our control and partly not ('natural'). In Willis's system, the last class would have been controlled by both cerebrum and cerebellum.

[113] For the cerebellum, see Kassler, *Music, Science, Philosophy*, pp. 1437.

[114] See Clarke and Jacyna, *Nineteenth-Century Origins of Neuroscientific Concepts*, p. 287.

[115] Since Fig. III.4 represents a side view of the pathway of the medullar trunk, anatomical substrates that are composed of two or more bodies show only half the number of which they are composed.

[116] Anatomists referred to these protuberances (now called 'corpora quadrigemina') as 'nates' and 'testes', terms that Willis disliked, although he continued to use them; see *The Anatomy of the Brain* [1664 tr. 1681], pp. 106–10. Unlike Descartes, he located the pineal gland (vulgarly, penis) between these protuberances; and he concluded that the use of this gland, like other 'Kernels', is to receive and retain within it 'the serous humors deposited from the arterious blood, till the Veins being emptied, may sup them back'.

[117] For background on the corpus of knowledge about the brain between 1600 and 1665, see the brief but useful survey by Clarke, 'Brain Anatomy before Steno', pp. 27–32, who, pp. 32–3, mentions the names of those who made new advances between 1655 and 1665.

[118] Willis's term 'corpora striata', like that of Briggs, clearly indicates more than one body. Indeed, when he first introduced this mechanism in *The Anatomy of the Brain* [1664 tr. 1681], p. 59, he used the term 'lentiform prominences', which not only suggests two bodies but also, significantly for what is to follow, bodies having the shape of a convex lens.

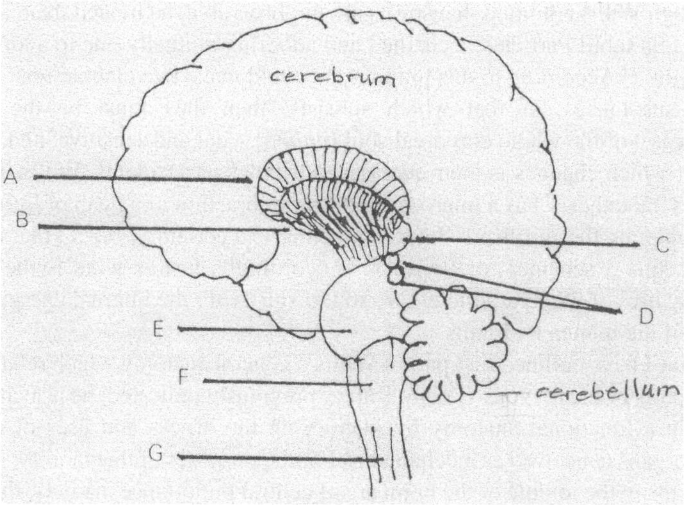

Fig. III.4 Upper part of the 'medullar Trunk' (side view) *after* the description in Willis, *The Anatomy of the Brain* (1664 tr. 1681)

A Corpus callosum
B One of a pair of corpora striata
C One of a pair of optic chambers
D Two of four 'orbicular protuberances' ('nates' and 'testes')
E Pons of Varioli
F Rest of the oblong marrow
G Spinal marrow (medulla spinalis) cut off

The medullar trunk, therefore, is not only a common trunk to which the cerebrum and cerebellum are attached, but also a common trunk to which all the nerves are attached and, hence, like a windchest of an organ, it receives from either the cerebrum or the cerebellum the animal spirits to be blown into this or that nerve.[119] As for the nerves themselves, although Willis sometimes described them as cords, strings or threads, yet he continued to hold the ancient opinion that the containers in the body form a network of *poroi*, that is, channels.[120] Consequently, the nerves, including their fibres, allow animal spirits to flow through these channels like wind in organ pipes.

---

[119] See Willis, *The Anatomy of the Brain* [1664 tr. 1681], pp. 122, 124, 126, 127, 130.

[120] See ibid., p. 127: 'The nerves themselves (as may be discovered by the help of a Microcosm or Perspective-glass) are furnished throughout with pores and passages, as it were so many little holes in a Honey-comb, thickly set, made hollow, and contiguous one by another; so the Tube-like substance of them, like an *Indian* Cane, is everywhere porous and pervious'. For a useful survey of early-modern nerve theories, see Clarke, 'The Doctrine of the Hollow Nerve', who, classed porosity theories as a variant of the hollow-nerve theory, included Willis's theory in this class and stated that Willis was probably the first to examine the nerve with a microscope. Newton is not mentioned in his survey, but I have already pointed out in Part II that his theory of compound matter, which includes the nerves, is a porosity theory and, hence, a variant of the hollow-nerve theory.

Although Willis admitted that spirits are unobservable, he treated them as a heap of 'Atoms or subtil Particles ... chained and adhering mutually one to another' like rays of light.[121] According to this premiss, he could make two claims: first, if spirits are real substances, or that which subsists, then they must be the essence ('hypostasis') of the whole corporeal soul (*anima*), vital and sensitive; and, second, since that which changes is their qualities, attributes or accidents, Willis imagined mutation ('metathesis') as a micro-mechanical interaction of a heap of little atomic bodies, whereby the spirits are 'figured together in a certain Species', that is to say, in 'phantasms', 'schemes' or 'types'.[122] Accordingly, just as wind is the internal energy or 'life' of the hydraulic organ, so too spirits are the internal energy or psychic life of the human hydraulis.

Thus far I have outlined that part of Willis's general anatomy which relates to the intrinsic parts of the nervous system. But, as previously indicated, he also attempted to provide a functional anatomy by identifying the offices and uses of different kinds of organs conceived as mechanisms. For the purposes of this inquiry, two such mechanisms in the middle of the brain are of central importance, namely, the corpus callosum and the corpora striata.[123] For he conceived the latter as a dark 'Chamber', the interior space of which is 'glassed with dioptric Looking-Glasses',[124] because 'in the Penetralia or inmost parts' of it,

> ...the Images or Pictures of all sensible things, being ... intromitted by the Passages of the Nerves, ... pass first of all through the streaked Bodies [corpora striata] as it were an objective Glass,[125] and then they are represented upon the Callous Body [corpus callosum], as it were upon a white Wall, and so induce a [sensory] Perception, and a certain Imagination of the thing felt.[126]

As will become evident below, Willis here described the interior space of the corpora striata as a mechanism that is isomorphic with the interior space of the eye, both of which are modelled on the interior space of a dark chamber with a white

---

[121] See Willis, *The Anatomy of the Brain* [1664 tr. 1681], p. 132; see also Willis, *Two Discourses* [1672 tr. 1683], p. 24.

[122] It is possible that Willis retained the scholastic doctrine of species, according to which movements like eddies represent the action of the transmuted 'power' of percussion, such action, though conceived as incorporeal, invisible and spiritual, by moving corporeal 'instruments' at any time becomes 'material'. For this doctrine, see Kassler, *The Beginnings of the Modern Philosophy of Music*, pp. 7–8, 10; for Thomas Hobbes's criticism of it, see Kassler, *Inner Music*, p. 50.

[123] Some commentators have substituted the term 'corpus striatum', or simply 'striatum' for Willis's term 'corpora striata'; see, e.g., Meyer and Hierons, 'A Note on Thomas Willis' Views on the Corpus Striatum'. Unfortunately, Connolly, 'Newton and God's Sensorium', p. 187, adopted their usage. For the meaning of this substituted term, see *infra* Conclusion, pp. 137–8, where it will be clear that the modern concept is not the same as Willis's concept, and, therefore, it is historically misleading to substitute the former for the latter.

[124] I.e., mirrors.

[125] I.e., object glass. See OED: The lens or combination of lenses in a telescope, microscope, or other optical instrument, which is situated nearest to the object, and thus receives the rays of light directly from it (1665–1839).

[126] Willis, *Two Discourses* [1672 tr. 1683], p. 25, see also pp. 33, 59, *et passim*.

wall on the posterior of the chamber. The only difference between them is that in the anterior of the chamber, the former mechanism includes a set of dioptric mirrors, whereas the latter mechanism includes a set of dioptric lenses.

Note, however, that in the passage quoted above, the images[b] are represented *upon* the corpus callosum, which suggests that Willis conceived its coat or *outer surface* as the white wall.[127] But the corpus callosum is also the seat for 'the thing felt' on its coat, for it is that feeling which induces a sensory 'Perception, and a certain Imagination' *in the under surface of* the corpus callosum, where 'all the Conceptions, Ideas, Forces, and Powers whatsoever both of the Rational and Sensitive Soul[s] are framed; and having there gotten a species and form, are produced into act'.[128] Accordingly, *in* the corpus callosum, the superior, immovable soul (*animus*) 'sustains the part of the Musitian playing on' the human hydraulis; for by 'governing and directing the animal spirits', it 'disposes and orders at its pleasure, the Faculties of the Inferior [movable] soul' (*anima*).[129]

What, then, is the use of the corpora striata? When dissecting these bodies through their middle, Willis noticed that they were marked with 'medullar streaks, as it were rays or beams' and that no other part is found 'chamfered or streaked after like manner'.[130] He therefore interpreted the striations as nerve-like channels for the animal spirits to pass from, or return to the interior space of the corpus callosum, where they circulated until called into act.[131] Hence, if the use of the corpora striata be inquired into, then as the striated bodies 'receive the forces of all the [external] Senses', so also [they receive] the first instincts[132] of spontaneous [i.e., voluntary] local motions'. For

> ...every influence from the Brain into the nervous stock, and on the contrary, from this [stock] into that [the brain], and the mutual commerce of the animal Spirits, must of necessity pass through these bodies [the corpora striata]. Wherefore all the [cranial] Nerves ... arise behind these chamfered bodies.[133]

In other words, the anatomical substrate variously called 'corpora striata', 'striated bodies' or 'chamfered bodies' is the seat of common sense, as well as the place of both passive-sensory and active-motor function, the psychosomatic process of which may be summarised as follows.

---

[127] But Willis, *Two Discourses* [1672 tr. 1683], p. 59, seems to have been uncertain whether images[b] transmitted 'into the streaked Bodies' are 'there represented as upon a white wall; ... or perhaps carried forward beyond into the Callous Body'.

[128] Willis, *The Anatomy of the Brain* [1664 tr. 1681], p. 77.

[129] Willis, *Two Discourses* [1672 tr. 1683], p. 34.

[130] Willis, *The Anatomy of the Brain* [1664 tr. 1681], p. 102.

[131] Ibid., p. 162. See also Willis, *Two Discourses* [1672 tr. 1683], p. 27, which is part of an 'accurate' description and use of the corpora striata with reference to the explanatory Table 8 and the accompanying figure of the human brain, pp. 26–9, 44.

[132] OED *sb.* 1 obs.: prompting, impulse = Instigation OED b: an incentive, stimulus, spur (1526–1869).

[133] Willis, *The Anatomy of the Brain* [1664 tr. 1681], p. 102.

From each peripheral end-organ ('sensory'), natural-extrinsic micro-mechanical motions are transmitted through the pores in the fibres of sensory nerves into the corpora striata, conceived as an optical instrument with a set of refracting or dioptric mirrors ('Looking-Glasses') whereby images[b] are represented as phantasms ('Species') on a white wall, the corpus callosum. If a phantasm imports mere apprehension of an object, the representation is sent for storage from the corpus callosum to 'cells' in the cerebral cortex, the anatomical substrate of (voluntary) memory. But if, from sensation or memory, the phantasm imports either good or evil, then it is reflected back by animal impulse from the corpus callosum through the corpora striata into the medullar trunk, from which the animal spirits are admitted to one or more of the motor nerve pipes and fibres implanted in the muscles for the appropriate motor response, so that good or evil may be embraced or removed.

The psychosomatic process just described is mediated by the knowing powers of the two souls, both of which are seated in the corpus callosum. In the case of animal impulse, the two powers, phantasie and intellect, mediate, respectively, the faculties of sensitive appetite and rational will, as may be illustrated briefly by recourse to Willis's identification of three ways—physical, moral and metaphysical—in which good and evil are brought to a person.[134] For either they respect the corporeal soul by herself, abstracted from any other relation; or they respect her as joined to the body; or they respect her as subdued by the incorporeal soul. But these ways, though separately identified, are often intertwined.

For example, because the corporeal soul is 'nearer to the Body', phantasie has an 'Affinity towards the Flesh' and is 'tied wholly to look to its profit and Conservation'.[135] Hence, if, during the passive-sensory part of the process, the phantasms are corrupted by intrinsic (vital) and extrinsic (natural) impulses, then the corporeal soul will incline to sensual pleasures and pains either by extending herself beyond 'the bounds of the Body' or contracting herself 'more narrowly' from the sphere of her 'Emanations'. And from 'this twofold Affection ... all the other Passions take their Origine', natural as well as preternatural.[136] Accordingly, phantasie symbolises the inconstant, undisciplined unregenerate person (*anima*) that requires the aid of the incorporeal soul (*animus*),[137] which exercises its knowing power of intellect to moderate and govern 'all the faculties and Acts' of the corporeal soul.[138]

---

[134] Willis, *Two Discourses* [1672 tr. 1683], pp. 46–8.

[135] Willis, *Two Discourses* [1672 tr. 1683], pp. 42–3.

[136] Ibid., p. 48. For the preternatural passions that perturb the nervous system, see especially Willis, *An Essay of the Pathology of the Brain,* [1667 tr. 1684].

[137] I.e., in the allegorical interpretation of the 'Fall' of man, phantasie represents Eve, whose property is sensitivity (*aisthēsis*); see p. 77 n.88 above. Note that as late as 1909, Upton, *Women in Music*, pp. 30–1, used the distinction between intellect and phantasie in order to distinguish the attributes of male and (detrimentally) female biological 'organisation' that, according to him, determine their respective roles in music.

[138] Willis, *Two Discourses* [1672 tr. 1683], p. 39. Cf. Briggs pp. 71–2 above.

Although intellect 'not only eminently Contains every Virtue' or operative influence of phantasie, yet, from images[b] represented on the coat of the corpus callosum, intellect 'deduces many other thoughts altogether unknown to the sense' and which phantasie, by herself, 'could no way Imagine'.[139] For intellect 'conceives the formal notions of Corporeal things, abstracted from all matter, and attributes to them prædicates[140] merely Intentional[,] yea and understands axioms or first principles alone, and as it were by a proper Instinct,[141] without recourse to Corporeal Species'.[142] Furthermore, intellect also, 'as it were by its own proper light', takes in or 'comprehends spiritual notions such as 'God to be Infinite and Eternal, that he ought to be Worshipped, that Angels or Spirits do inhabit the World ... and many other notions merely Spiritual, by no means to be learnt from Sense or Phantasie'.[143] Intellect, therefore, symbolises the normative person, which Willis described as 'the supream and principal form of the whole Man.[144]

Seated during earthly life in the interior space of the corpus callosum, the incorporeal soul (*animus*) exercises its power of intellect, which '*beholds* the Images and Impressions represented by the sensitive Soul [*anima*], as in a looking Glass,[145] and according to the Conceptions and notions drawn from thence, exercises the Acts of Reason, Judgment, and Will'.[146] It also '*beholds* all enunciations' conceived by Phantasie' and then 'judges them, whether they be true or false'. What is more, it '*beholds* it self, by a reflected Action',[147] and, hence, has self-perception. For it contemplates, or reflects on either its own logical operations or its own light. But none of its perceptions are the result of corporeal percussions of the animal spirits, because the cognitive perceptions of intellect are 'pure', that is to say, incorporeal. Nevertheless, if what is represented to it from a brain 'evilly affected, and as it were monsters from a multiplying or distorted glass', then 'for that reason Intellect

---

[139] Ibid. For 'the sense', read here and also below 'the sensitive part of the corporeal soul'.

[140] I.e., in logic that which is predicated or said of the subject of a proposition; the second term of a proposition, which is affirmed or denied of the first term by means of a copula.

[141] OED b: Immediate apprehension by the intellect alone (1659–1877).

[142] Willis, *Two Discourses* [1672 tr. 1683], p. 39. In the passage quoted he reveals his indebtedness to the conception of intelligence (i.e., knowledge) of Platonic tradition. In Willis's version, the incorporeal soul (*animus*) exercises its power of intellect in the acquisition of intelligible truths. Accordingly, knowing is an intuitive grasp of such truths and thinking is a logical progression towards them. Cf. the version of the Renaissance philosopher, Ficino, 'Five Questions concerning Mind', pp. 197–207.

[143] Ibid., referring to divine truths, which during the Middle Ages were conceived as exemplar ideas illuminated not by the human intellect but by God; see Robinson, *The Shape of Things Known*, p. 34.

[144] Willis, *The Anatomy of the Brain* [1664 tr. 1681], 'Preface to the Reader'. This conception may imply that the Musician (see p. 86 above), seated in its organ room (the corpus callosum), actually represents that part of the human hydraulis on which it plays, i.e., the part concerned with voluntary function. For a later, fully developed representative theory based on a musical-instrument model, see Kassler, *Music, Science, Philosophy*, pp. 286–90.

[145] I.e., mirror.

[146] See Willis, *The Anatomy of the Brain* [1664 tr. 1681], p. 32 (italics mine).

[147] Ibid., p. 39 (italics mine).

beholds or forms conceptions and thoughts only deformed, distracted one from another, and very confused'.[148]

In the passages quoted in the previous paragraph, the repeated use of the word 'behold' suggests that Willis conceived intellect not only as an eye but also as an emblem of the incorporeal soul. Seated in the corpus callosum, this soul, exercising its power of intellect, beholds the spectacle in the space 'out there' not as the things themselves but as corporeal images[b] transmitted through the striations in the corpora striata and represented on the coat of the corpus callosum. Accordingly, in the common sensory, understood in Willis's enlarged sense, the incorporeal soul is shut up not only with its own sensory perceptions, but also with its own reflections.

3.3 The Spectator in the Dark Room

In conceiving the knowing power of the incorporeal soul as an intellectual eye, Willis privileged sight above the other external senses; and he gave the following reason for doing so:

> If there be any strife for Dignity among the Senses, the Palm is given, almost by the consent of all, to Seeing, as the most noble Power; because this faculty apprehends things at a great distance, under a most subtil Figure, by a most clear perception, and with great delight; so *this Sense acts* ... [in a way] *that is next in virtue to the Eternal and Immaterial Soul.* To wit, it views and measures both Heaven and Earth in a Moment, and brings within its embraces whatever Bodies are situated in either, and that are far remote from our touch[149]

That sight is the best of the senses was a commonplace with roots in antiquity. But in the passage italicised above, Willis also provided a hint that the eye is the window of the soul; and over the centuries this commonplace was used to support the 'Dignity' of sight.[150] Yet other arguments were also used to privilege sight.

For example, although Plato singled out for consideration the so-called 'intellectual' senses, seeing and hearing, yet he gave the palm to the former, because 'the cause and purpose of god's invention and gift to us of sight was that we should see the revolutions of intelligence in the heavens and use their untroubled course to guide the troubled revolutions in our own understanding, which are akin to them'.[151] Like Plato, Aristotle singled out seeing and hearing. But he offered a different reason for giving the palm to sight by claiming that it is more important not only 'in itself' but

---

[148] Willis, *Two Discourses* [1672 tr. 1683], p. 179 (italics mine); see also pp. 188, 209–10, for distorted images[b] due to brains 'evilly affected' or malformed.

[149] Willis, *Two Discourses* [1672 tr. 1683], p. 75 (italics mine); see also p. 82: the eye, being round, imitates 'the System of the World'.

[150] See Frangenberg, '*Auditus visu prestantior*', p. 73 and n.8.

[151] Plato, *Timaeus*, 47. Like sight, the gift of hearing was given to us 'for the sake of harmony, which has motions akin to the orbits in our own soul'. But the lesser position of hearing is accounted for by Plato's warning against using 'audible musical sound' to give 'irrational pleasure' rather than to reduce 'to order and harmony any disharmony in the revolutions within us'.

also 'for the mere necessities of life'. For 'sight informs us of many differences of all kinds, because all bodies have a share of colour, so that it is chiefly by this medium that we perceive the common sensibles', namely, 'shape, magnitude, movement and number'.[152] Indeed, for Aristotle, 'sight is paradigmatically sense-perception'.[153]

In Newton's circle of acquaintances, at least three men propounded the dogma that sight is the primary sense. Moreover, all three men had either direct or indirect acquaintance with Willis's common sensory concept. Since William Briggs's contact with Newton, as well as his indebtedness to Willis has already been noticed,[154] it remains to point out that at the outset of his 1675 treatise, he presented a brief argument from design concerning '*Occuli præstantia*' (pre-eminence of the eye). In sum, his argument was that the eyes display the wisdom of 'the supreme Deity', as well as the honour he bestowed upon them. For '*no other part* embraces so divine an ingenuity and order, *nor is any part* constructed with such a huge apparatus of vessels and membranes'.[155]

Of Newton's two other acquaintances who privileged sight, one has been mentioned briefly in Parts I and II of this inquiry, namely, Robert Boyle,[156] who in the autumn of 1655 took up lodgings in Oxford at Deep Hall, the High Street home of John Cross, a near neighbour of Willis and one of two apothecaries that Willis dealt with at this time.[157] About the same time, Boyle became a member of an Oxford Club in which Willis was already a leading figure.[158] In addition to conducting chemical experiments, this Club held meetings for the purposes of discussion and

---

[152] Hett (tr.), *Aristotle On the Soul*, 'Parva naturalia', pp. 218–9, where Aristotle added that 'for the mind and indirectly[,] hearing is the more important', indirectly because 'hearing makes the largest contribution to wisdom'.

[153] See Aristotle, *De Anima*, III.iii. in Lawson-Tancred (tr.), *Aristotle De Anima*, p. 200. Cf. Hett (tr.), *Aristotle On the Soul*, pp. 162–3 ('sight is the chief sense'). See also Crowther and Barker, 'Training the Intelligent Eye', p. 436, for the privileging of sight in Aristotle's *Metaphysics*.

[154] See p. 61 above. As indicated there, Briggs first came into contact with Newton in the 1660s, when the two men were at Cambridge University. As for Willis, whose influence on Briggs was there evidenced, his professional life began in Oxford and ended in London, so that it is unlikely that the two men were directly acquainted, because Briggs was in France from c.1669 and returned to Cambridge in 1672; and three years later Willis died in London at his house in St. Martin's Lane.

[155] Briggs, *Opthalamo-graphia* [1675], Ch. 1, sect. 1, pp. 1–3; cf. Briggs, *Opthalamo-graphia* [1685 tr. 1990], Ch. 1, sect 1, pp. 1–2 (italics mine).

[156] According to Westfall, *Never at Rest*, pp. 288–9, Newton and Boyle first met in 1675. For Boyle's publications owned by Newton, or accessible to him through Barrow, see Harrison, *The Library of Isaac Newton*, (Nos. 254–81), pp. 107–9, and Feingold, 'Isaac Barrow's Library', (Nos. 172–83), pp. 345–6.

[157] See Dewhurst, *Willis's Oxford Lectures*, p. 8 and Plate 8. Note that Cross was also a friend of the Oxonian, John Fell, a college contemporary of Willis, who later married Fell's sister.

[158] See ibid., pp. 1–36. The Club, which met between c.1648 and 1667, used the laboratory facilities at Wadham College until 1659; but after Boyle set up his first laboratory in Deep Hall, the Club's meetings were held there. According to Frank, Jr., *Harvey and the Oxford Physiologists*, pp. 183 and nn.160–2, two divines on the periphery of the Club followed the anatomical discussions closely, namely, Samuel Parker and Robert Sharrock. In their publications the former, relying on Willis's anatomy, presented two arguments, one from design, the other for the pre-existence of souls, whereas the latter included a digression on Willis's conception of the use of the corpora striata and corpus callosum.

debate, one of its topics being the problem of muscle action.[159] Between 1661 and 1664 when Boyle was occasionally absent from his Oxford lodgings, Richard Lower, Willis's assistant and colleague, kept him informed concerning Willis's activities, including his dissections and experiments, his progress towards publication of *Cerebri anatome* and his Sedleian lectures on natural philosophy,[160] which, with a brief hiatus during the plague, were read between c.1660 and 1667.[161]

Although Willis's name occurs only once in Boyle's collected writings,[162] yet those writings provide evidence that Boyle not only privileged sight but also relied on a number of Willis's main concepts. For example, he produced several arguments from design concerning the eye, to which he devoted more attention than any other external sense.[163] Moreover, he held that sight had 'the greatest cognation with the understanding',[164] which is 'to be looked upon as the eye of the mind'[165] and as 'the highest faculty' in humans.[166] And he adopted Willis's general anatomy and its musical-instrument model not only by describing the living body as an 'hydraulico-pneumatical engine',[167] but also by identifying all three mechanisms of Willis's control system, namely, the cerebrum, cerebellum and medullar trunk (that is, the oblong marrow and its continuation, the spinal marrow).[168] Accordingly, he constituted the living engine of 'solid and stable parts' (the containers), as well as observable and unobservable fluids (the contained substances), the latter of which, he claimed, are in 'organical motion'.[169]

---

[159] According to Bennett, 'A Note', pp. 63–4 and n.30, one of the members of the Club, Christopher Wren, in an address to the Royal Society read about 1662, pointed out that a mechanical account of the muscles must be complemented by a chemical explanation; and in March 1665, Wren himself made an experiment to explain muscle action by 'explosion'. It should be noted, therefore, that Willis, *The Anatomy of the Brain* [1664 tr. 1681], 'Preface to the Reader', acknowledged Wren not only for delineating '*many Figures of the Brain and Skull*' reproduced in his book, but also, with Thomas Millington (another member of the Club), for being 'frequently' present '*at our Dissections, and to confer and reason about the uses of the Parts*'.

[160] See Boyle, *The Works*, vol. 6, pp. 462–81; for Willis, see also Boyle, *The Correspondence*, vols. 1–4.

[161] See Dewhurst, *Willis's Oxford Lectures*, pp. 37–49. In 1667 Willis and in 1668 Boyle left Oxford for London.

[162] See Boyle, *The Works*, vol. 4, p. 503, where, in a convoluted sentence, he indicated that he used the term 'chymical principles' in Willis's 'enlarged' sense, i.e., to mean five, not three principles, as noted p. 76 and n.85 above.

[163] See ibid., vol. 6, for the invaluable analytical index (not paginated).

[164] I.e., the intellect, ibid., vol. 5, p. 242.

[165] Ibid., vol. 4, p. 461.

[166] Ibid., vol. 2, p. 6.

[167] Ibid., vol. 4, p. 442; vol. 5, p. 232.

[168] Ibid., vol. 6, p. 740.

[169] Ibid., vol. 4, p. 442; vol. 5, p. 232. The observable fluids include the blood and 'liquors', the latter of which are so disposed 'by things analogous to local ferments' as to be put into 'a fermentation or commotion'. For there are 'chymical uses' of 'some parts, that serve for the chemical elaboration of spirits and other fluids'.

In addition, Boyle's method of physiological explanation follows the Galenic tradition of interpretation, described previously,[170] for he posited two disposing principles, the first of which is 'the human mind [intellect], or rational soul' (*animus*),[171] conceived as 'an immaterial and immortal being that bears the image of its divine maker'.[172] Residing in an undisclosed place in the anatomical space of the brain, this soul exercises its power of intellect to judge 'wherein its own nature consists, and whether or no itself be a distinct faculty from the will'.

> And to come yet closer to the point, ... logic and metaphysics are the works of the human intellect, which by framing those disciplines, manifests, that it does not only judge of ratiocinations, but of the very principles and laws of reasoning... To these things it is agreeable, that *if we will compare the bodily eye with the understanding, which is the eye of the mind*, we must allow this difference, that the intellect is as well a looking-glass as a sensory, since it does not only see other things, but itself too, and can discern its own blemishes.....[173]

In the passages quoted above, Boyle muddled up soul, mind and intellect, a muddle to be met with later on. For initially he equated mind and soul; then he made intellect into '*the eye of the mind*', equating that intellectual eye with a mirror ('looking glass'), as well as a central end-organ ('sensory'). But if intellect is an eye that sees in a mirror the images[b] of things, as well as itself, then those images would have to be cast upon, or reflected from a corporeal surface in or on the common sensory.[174] Only then, when illumined 'by the evidence of [sensory] perception', can intellect judge 'other things', as well as 'itself'.[175]

Boyle's second disposing principle, the corporeal soul (*anima*),[176] has three faculties—common sense, imagination and locomotion. The first faculty '*both* discerns those objects of sense that are common ... *and* distinguishes the proper objects of one particular sense from that of another'.[177] The second faculty, which is 'most of kin to intellect',[178] 'exhibits the images as it were of sensible things, which are variously ... altered', because imagination 'often dresses up such a large variety of corporeal *ideas* and fantasms ... as makes the fancy a most strange and many times oddly furnished *scene*' that surpasses those devised 'for theatres, operas, and masks'.[179] But the third faculty 'alone may suffice to challenge our wonder', because

---

[170] See *supra* Pt.II.2.2, pp. 52–3.

[171] Boyle, *The Works*, vol. 2, p. 731.

[172] Ibid., vol. 4, pp. 19–20.

[173] Ibid., vol. 4, p. 419 (italics mine).

[174] See OED: Species, *sb.* I. 3c, obs.: The image of something as cast upon, or reflected from, a surface; a reflection (1638–1737).

[175] Boyle, *The Works*, vol. 4, pp.460–1.

[176] Ibid., vol. 6, p. 740. For this disposing principle as the whole of a distributed body of animal spirits in the brain and nervous system, see ibid., vol. 4, pp. 442, 457; vol. 5, p. 10; vol. 6, pp. 739–40.

[177] Ibid., vol. 6, p. 741.

[178] Ibid., vol. 4, p. 415.

[179] Ibid.

> ...such minute and invisible corpuscles as animal spirits, and these too, in no greater numbers than can be furnished and contained in the brain and inconspicuous cavities of the nerves, *even supposing those parts assisted by some small portion of liquor to make explosions*, should be able to move all the outward limbs and internal muscles of the body.[180]

As the italicised portion in the above passage reveals, the supposition in question is Willis's pre-1670 explanation of the motive power or unknown material cause that swells the muscle on contraction.[181]

Finally, there is Newton's acquaintance, as well as Willis's medical colleague at Oxford, John Locke,[182] who described sight as

> ...the most comprehensive of all our Senses, conveying to our Minds the *Ideas* of Light and Colours, which are peculiar to that Sense; and also the far different *Ideas* of Space, Figure and Motion, the several varieties whereof change the appearance of its proper Object, *viz.* Light and Colours, we bring ourselves by use, to judge of the one by the other.[183]

Accordingly, for sensory perception he relied on such terms as 'see', 'view', 'look at', 'observe' and their cognates, thereby identifying intentional life (consciousness) with seeing, hence assuming that '*Ideas*' function like pictures in representing their objects.[184] In addition, he acknowledged that the 'Perception of the Mind ... [is] most aptly explained by words relating to the Sight',[185] and, therefore, for cognitive perception he identified 'the perceptive Faculty of the Mind' as the eye of the understanding, which he described in such terms as quick-sighted, short-sighted or blinded, and as possessing foresight and insight, though apparently not hindsight.[186]

It should be noted, therefore, that the tradition in which sight is conceived as the basic sense also leads to epistemologies indebted to the semantic field of vision,[187] of which Locke's *Essay concerning Human Understanding* is a notable example. Indeed, in the words of his modern editor, P. H. Niditch, Locke assumed that by a process of self-examination, it would be possible to discover 'an adequate ophthalmology of the eye of the understanding'.[188] According to Niditch, the term 'understanding' corresponds to the Latin term 'intellectus'. And this is indisputable. But

---

180 Ibid., vol. 6, p. 743 (italics mine).

181 See *supra* Pt.II.2.1, pp. 43–5.

182 According to Westfall, *Never at Rest*, pp. 289, 488, contact between Locke and Newton began in the late 1680s, probably 1689.

183 Locke, *An Essay concerning Human Understanding*, p. 146; see also, p. 43.

184 This assumption has its roots in Aristotle's theory that the soul perceives an image 'in its own right' and a copy of the image as a reminder; see Sorabji (tr.), *Aristotle on Memory*, pp. 47–50, especially pp. 48–9, 51–2.

185 See Locke, *An Essay concerning Human Understanding*, p. 363.

186 See ibid., pp. 52, 127, 131, 136, 144, 149–50, 209, 226, *et passim*.

187 For a brief but useful history of visual epistemology from Parmenides to Pomponazzi, see Robinson, *The Shape of Things Known*, pp. 11–59, according to whom, pp. 36–7, the two main traditions of it derive from Plato and Aristotle until these traditions were fused by Boethius, after which his fusion was to become the norm in medieval philosophy.

188 See Niditch in Locke *An Essay concerning Human Understanding*, p. xix.

he also claimed that Locke 'usually' applied the term 'mind' to represent 'only the understanding'.[189] And this claim is disputable. For Locke's use of the terms 'soul', 'mind' and 'understanding' (intellect) is inconsistent (for some instances see further below). And this inconsistency leads to an ambiguity in Locke's terminology, an ambiguity that David Hamlyn described as 'not without significance', because 'it reflects an ambiguity of thought'.[190]

In 1969 Gerd Buchdahl claimed, without demonstrating, that Newton framed his conception of the human *sensorium* 'in terms which borrow considerably from Locke's theory of perception'.[191] Unknown to Buchdahl, however, Locke owed a debt to Willis, even though he never named that physician in the *Essay*. To illustrate this indebtedness, it should be noted that Locke believed that '*Ideas* in the Understanding, are coeval with Sensation'.[192] Consequently, it was necessary for him to provide some anatomical details concerning how this or that 'motion must be continued [from the external senses] by our Nerves, or animal Spirits, by some parts of our Bodies, to the Brains [*sic*] or the seat of Sensation, there to *produce in our Minds the particular* Ideas *we have of them*'.[193]

According to Locke, the nerves are merely 'Windows' that convey motions 'from without to their Audience in the Brain, the mind's Presence-room (as I may so call it)', where those motions are 'perceived by the Understanding'.[194] So that 'external and internal Sensation are the only passages that I can find, of Knowledge, to the Understanding', and that: 'These 'alone [i.e. external and internal sensation and not the nerves], as far as I can discover, are the Windows by which light is let into this *dark Room*'. For

> ...the *Understanding* is not much unlike a Closet wholly shut from light, with only some openings left, to let in external visible Resemblances, or *Ideas* of things without, [and] would the Pictures coming into such a dark Room but stay there, and lie so orderly as to be found upon occasion, it would very much resemble the Understanding of a Man, in reference to all Objects of sight, and the *Ideas* of them.[195]

In the several passages quoted above, Locke first described the nerves as conduits ('Windows') to the seat of sensation, afterwards substituting for the nerves 'external and internal Sensation'. He described the seat itself as 'the mind's Presence-room', but he failed to disclose the specific place of that seat in the anatomical space of the brain. If by the term 'mind' he consistently denoted the understanding, as his mod-

[189] Ibid., p. xxii.

[190] See Hamlyn, *Sensation and Perception*, pp. 96–7.

[191] Buchdahl, *Metaphysics and the Philosophy of Science*, p. 596. For a brief critique Buchdahl's claim, see *infra* Conclusion pp. 127–8. For critical assessment of Locke's theory of perception, see Hamlyn, *Sensation and Perception*, pp. 94–104.

[192] See Locke, *An Essay concerning Human Understanding*, p. 117. Note that Newton owned the first (1690) edition of this book, as well as a Latin translation of the fourth (1701) edition; see Harrison, *The Library of Isaac Newton*, (Nos. 967, 698), p. 181.

[193] See Locke, *An Essay concerning Human Understanding*, p. 136.

[194] Ibid., pp. 121.

[195] Ibid., pp. 162–3.

ern editor stated, then does not Locke's own statement—that 'the *Understanding* is not much unlike a Closet'— imply that the understanding is an anatomical substrate or mechanism, not a function?

However, these and other muddles were not from ignorance, since during his residence in Oxford he read Willis's *Cerebri anatome*,[196] as well as attended, and took notes on Willis's Sedleian lectures.[197] Indeed, some thirty four of those lectures have been preserved in one of Locke's notebooks,[198] including a lecture on the cerebrum,[199] in which Willis made the following statements:

> In the corpus callosum are attached the corpora striata connecting the cerebrum to the legs of the medulla oblongata.[200] In these corpora there are some striæ passing upwards and others downwards and through them the spirits and images of sensible things pass from the medulla oblongata into the cerebrum, while spirits initiating movement descend into the medulla oblongata. These corpora [striata] seem to be **the primary sensorium**[201] by which images ... are first brought in and recorded; and from there proceed into the corpus callosum where imagination occurs....[202]

As indicated in the previous section, Willis treated sensory perception as a passion of the inferiour soul (*anima*), which occurs when images[a], formed on the coat of the retina, are afterwards recorded on the coat of the corpus callosum at the posterior of the common sensory. Whereas he treated cognitive perception as an action of the superior soul (*animus*), which occurs when images[b] on the coat of the corpus callosum are not only 'beheld', but also reflected upon by the soul's power of intellect. Accordingly, there are two sources of perception, one from sensation which is passive, the other from reflection which is active. If we emend the word 'images' (in both its senses) to ideas,[203] then we have the basis of Locke's theory of perception, which, in short, is that '*Ideas* got by *Sensation*', which 'for the most part' are 'only passive',[204] are caused when '*Impressions* ... are made on our *Senses* by outward Objects ... extrinsical to the Mind', whereas '*Ideas* of *Reflection*' are caused when 'the Mind comes to reflect on its own *Operations*'.[205]

---

[196] See Dewhurst, *John Locke*, pp. 11–2. Like Boyle, Locke utilised Willis's concepts for his own purposes; see, e.g., Locke *An Essay concerning Human Understanding*, pp. 484–5, where the 'liquor' in question is Willis's *succus nervosus*.

[197] At the termination of the lectures in 1667, both Willis and Locke left Oxford for London; see Dewhurst, *Willis's Oxford Lectures*, pp. 37–49.

[198] See ibid., p. viii, whose edition of the lectures is a collation of Locke's notes with those of Richard Lower. Note that the substance of some of the lectures afterwards appeared in expanded form in Willis, *Pathologiæ cerebri, et nervosi* [1667] and *De anima brutorum* [1672].

[199] See Dewhurst, *Willis's Oxford Lectures*, pp. 138–43 and n.1.

[200] I.e., upper part of Willis's medullar trunk (or wind chest); see Fig.III.4 above.

[201] OED Sensory *sb*. 2 = Sensorium, obs. as common or first sensory (1653–1829).

[202] Dewhurst, *Willis's Oxford Lectures*, p. 141 (bold face mine).

[203] I.e., Species; see OED *sb*. 6, obs. in Platonic philosophy = Idea *sb*. 1, obs. (1678–1792).

[204] Locke, *An Essay concerning Human Understanding*, p. 143.

[205] Ibid., pp. 117–8.

These latter ideas are so-called, because 'the Mind turns its view [i.e., eye] inwards upon itself, and contemplates its own Actions',[206] which Locke identified as perceiving and preferring, or in 'the ordinary way of Speaking', as understanding and willing. Since these *actions* are 'Powers of the Mind', he could describe them, respectively, as 'Thinking and Motion'.[207] But from his reliance on the semantic field of vision, it is clear that he conceived the understanding, substantively, not as the mind, a turning mechanism, but as 'the Eye of the Soul',[208] the soul, of course, being what Locke had referred to earlier as the 'Audience in ... the mind's Presence-room' or dark chamber. Accordingly, the soul (*animus*), and not the eye of the understanding, is shut up with its own ideas, because somewhere within the interior space of the '*dark Room*', a spectator perceives the spectacle in the space 'out there' as representative ideas. In terms of epistemology, therefore, Willis and Locke share the same externalist premiss, namely, that knowledge is determined by external objects.[209] And this, too, is the premiss implicit in Newton's texts.[210] Accordingly, these three men share the central tenet of psychological empiricism[211]—that all knowledge must ultimately be derived from, and relate to the empirical world, the world of which we have experience.

---

[206] Ibid., p. 226. The various meanings of Willis's term 'behold' are now obsolete, and Locke did not employ the term. Instead, he used the term 'contemplate' and its synonyms; see OED 1. *trans.*: To look at with continued attention; gaze upon, view, observe.

[207] Locke, *An Essay concerning Human Understanding*, pp. 236–7.

[208] Ibid., p. 152: In contemplating or searching for some idea 'lodg'd *in* the *Memory*', 'The Mind ... turns, as it were, the Eye of the Soul upon it'.

[209] Like Willis p. 87 n.142 above. Locke retained the Platonic tradition of knowledge as information (intelligence). For the former's statement that the cerebrum at birth is a '*tabula rasa*', see Dewhurst, *Willis's Oxford Lectures*, p. 66; and for the latter's statement that the 'yet empty Cabinet', like 'white Paper void of all Characters', needs to be furnished with '*Ideas*' from both sensation and reflection, see Locke, *An Essay concerning Human Understanding*, pp. 55, 81, 104.

[210] See *supra* Pt.I.1.2, Texts III, V, VI, VII (Ques. 12, 13), VIII, XI and XII.

[211] Locke is the main representative of this variety of empiricism; see Kassler, *Music, Science, Philosophy*, pp. 279–92.

Part IV

IV
Generalising to the Divine *Sensorium*

4.1 Taking Stock

Thus far I have focused on the human *sensorium* by attempting to discover in the first two parts what can be learned about it from Newton's texts and, then in the third part, from the wider context there treated. Recall that at the end of Part II,[1] four questions were posed that could not be answered from Newton's texts; and although the wider context provided insight into how these questions might be answered, it seems that none of the four questions can be answered with absolute certainty. Nevertheless, if we are to understand Newton's complete *sensorium* concept, answers are required for reasons that will become clear in the next section. Accordingly, before addressing the problem of the divine *sensorium*, each question will be considered here and a plausible and, insofar as possible, an evidence-based answer will be provided. But in attempting to provide such plausible answers, I will adopt a slightly different order from that in which the four questions were first posed.

(Question 1) Since Newton treats as receptors both the light-sensitive substance in the retina and the sensitive substance in the *sensorium*, does this mean that he also extended to the *sensorium* the *camera obscura* model that he had used to explain not only the functional anatomy of the eye, but also image formation on the coat of the retina?

To the best of my knowledge, Willis was the first anatomist to have extended the model of the interior space of the dark chamber to the common sensory, taken in his enlarged sense to include the corpora striata and the adjacent coat of the corpus callosum.[2] He thus treated the interior space of the common sensory as structurally similar to the interior space of the eye by conceiving both anatomical substrates as optical instruments with a white wall at their posterior part of each 'chamber'. The

[1] See *supra* Pt.II.2.2, p. 57.

[2] See *supra* Pt.III.3.2, Fig. III.4.

J. C. Kassler, *Newton's* Sensorium*: Anatomy of a Concept*, Archimedes 53,
https://doi.org/10.1007/978-3-319-72053-1_4

main difference between the two chambers is that the anterior part of the eye includes a set of lenses (dioptric glasses), whereas the anterior part of the common sensory includes a set of mirrors (dioptric looking-glasses).[3]

Nevertheless, by extending the *camera obscura* model to the common sensory, Willis could apply the optical processes of the model to the problem of how images[b] are formed on the coat or *exterior* surface of the corpus callosum from impressions felt by the sensitive substance within the *interior* surfaces of the corpora striata.[4] But he could not explain how the immaterial soul (*animus*) seated in the *interior* space of the corpus callosum exercises its intellectual eye in order to behold the spectacle represented on the coat of that container. The reason for this is that *animus*, a disposing principle, and its intellectual eye, a disposing power, are known only through their effects, that is to say, they are unknown causes. And, as indicated in Part II, this causal-analytical method of physiological explanation is traceable to Galen.[5]

In 1672, the year in which Willis made public his model of the common sensory, Newton announced his invention of a type of reflecting telescope where the spectator stands at right angles to the line of view. This optical instrument included 'two Metallin *speculum's*, the one Concave, (instead of an Object-glass) the other Plain', i.e., flat. These specula or mirrors were positioned in the optic tube of the telescope as follows:

> ...the Tube ... is open at one end which respects the object; ... the other end is close[d], where the said Concave [speculum] is laid, and ... near the open end there is a flat oval *speculum*, made as small as may be, the less to obstruct the entrance of the rays of Light, and inclined towards the upper part of the Tube, where is a little hole furnish't with the said Eye-glass.[6] So that the rays coming from the object, do first fall on the Concave [speculum] placed at the bottome of the Tube; and are thence reflected toward the other end of it, where they meet with the flat speculum, obliquity posited, by the reflection of which they are directed to the little plano-convex [eye-] Glass [i.e., ocular], and so to the spectators Eye, who looking downwards sees the Object, which the Telescope is turned to.[7]

It is important to note that Newton's type of reflecting telescope positions the glass of the ocular at the top end of the telescope and, hence, reverses the position of the anterior and posterior parts of Willis's common sensory. Bearing this in mind, the

[3] See *supra* Pt.III.3.2.

[4] See *supra* Pt.III.3.2, pp. 79, 85.

[5] See *supra* Pt.II.2.2, p. 84–5.

[6] I.e., the ocular or eyepiece.

[7] Newton, 'An accompt of a New Catadioptrical Telescope', reprinted in Cohen and Schofield (eds.), *Isaac Newton's Papers*, p. 61. The early reflecting telescopes were based on lenses, not mirrors; but in the hands of Galileo and Kepler, they brought about telescopic astronomy; see Van Helden, *Measuring the Universe*. In the epilogue to his *Optica promota* (1663), James Gregory proposed a new type of reflecting telescope that is put to the eye and pointed straight at the object to be viewed. It included a concave objective mirror and a concave secondary mirror that reflects the light from the object glass through a small hole in the middle of the objective mirror into the ocular. In the same year Gregory commissioned a London optician to construct a telescope according to his design; and in 1674 Robert Hooke, who had 'direct experience of the very type of reflecting optics that Newton had harnessed', presented the first Gregorian telescope to the Royal Society; see Simpson, 'Robert Hooke and Practical Optics', pp. 44, 49. According to Newton, he got the idea of using mirrors from Gregory; see Newton, *The Correspondence*, vol. 1, p. 153.

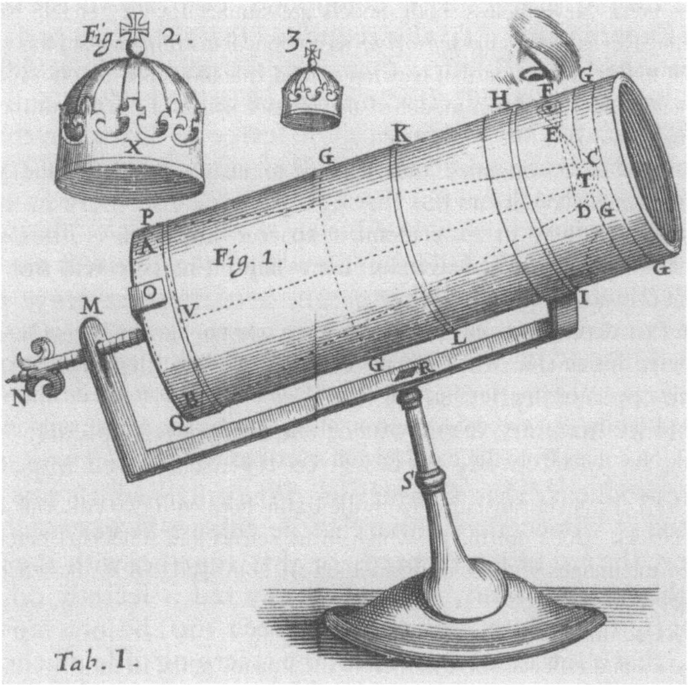

Fig. IV.1 Newton's catadioptrical telescope *from* a facsimile of the 1672 original in Cohen and Schofield (eds.), *Isaac Newton's Papers*, p. 60

AN	Concave speculum
CG	Flat oval ('plain') speculum
F	Small eye-glass for the spectator's eye

design of his reflecting telescope provides a clear indication of how the spectacle in the space 'out there' is represented to the spectator's eye. For the objective ('Concave') speculum collects rays of light at the bottom of the optic tube (the posterior part) and reflects them (upwards) to the flat speculum near the open end of the tube (its anterior part). Accordingly, the spectator's eye (looking downwards through the ocular) beholds a representation of the spectacle on the flat speculum and not the spectacle itself (see Fig. IV.1).

If we take Newton's description of the reflecting telescope into account, then, to answer Question 1, it seems plausible to assume that he conceived the *sensorium* as an optical instrument with two mirrors, one of which (the concave objective speculum) functions as a reflector and the other (the flat speculum), as a receptor. Of course, there are differences in both structure and function between Newton's Keplerian model of the interior space of eye and his model of the interior space of *sensorium*. Structurally, the former model includes a refracting lens and a receptor, whereas the latter model includes a reflecting mirror and a receptor mirror. Functionally, the image formed on the Keplerian receptor is inverted, whereas the image formed on the Newtonian receptor is erect. Nevertheless, both models are similar, for they enable Newton to conceive the eye, as well as the *sensorium* as an optical instrument. And this similarity might lead to the conclusion that Newton extended analogically, but not like Willis, isomorphically, his Keplerian model of the interior space of the eye to the interior space of the *sensorium*.

If this conclusion is granted, then it helps to answer another of the questions posed at the conclusion of Part II:

(Question 4) Where in the anatomical space of the brain is the seat of the immovable soul that exerts power by impressing on its corporeal and movable instrument (*anima*)?

Figure IV.1 depicts not only Newton's telescope but also an eye of a spectator. Accordingly, this model, when extended to the middle of the brain, requires an immovable spectator, the intellectual eye of which looks down on the spectacle represented on the flat (receptor) mirror in the *sensorium*. Some support for this interpretation comes from three of Newton's texts as follows:

(1) Text VIII (English translation from the Latin, here with two editorial additions for clarity): 'Does not that[,] which in us senses and reflects[,] perceive and survey the images of those things, which are conveyed to it by the sense organs, into its little sensory'.
(2) Text IX comment (English translation from the Latin): 'certainly, that which in us senses and reflects, surveys merely the Images in the Brain'.
(3) Text XI (Newton's own translation and modification of Text VIII): when images[b] are 'carried' into the *sensorium*, they 'are there seen and beheld by that which in us perceives and thinks'.

Though differently expressed, the implications are the same in all these passages. For, first, the phrase, 'that which in us', implies the soul (*animus*). Second, this soul never senses or reflects, perceives or thinks, without exercising its power of intellect to see or behold the images[b] represented in the *sensorium*. Third, the soul's cognitive perception of images from sensation, as well as from reflection is always indirect. Therefore, to answer Question 4, in order for this soul to behold such images, it must be seated *above* the coat of the *sensorium*, so that its intellectual eye, by looking downwards from some undisclosed place in the anatomical space of the brain, beholds the spectacle represented to it on the receptor (flat mirror) at the anterior of the under surface of the *sensorium*.

And because, according to Text II, it is this soul, and not the corporeal soul (*anima*), which is the unknown efficient cause of voluntary actions,[8] we may ask:

(Question 3) Since the human *sensorium* is the seat of sensory perception, is it also the seat of locomotion?

Referring to the summary at the end of Part II concerning Newton's conception of sensory-motor function, it is clear from his texts that each function has a different anatomical pathway through which stimuli are transmitted. For vision, his paradigm of sensory function, the pathway is from the optic nerves and their fibres that exit

[8] See *supra* Pt.II.2.1, p. 38 and Pt.II.2.2, p. 48.

behind the retina of each eye and that terminate on either side of the *sensorium* in the brain. But since there are diverse kinds of animal motion, he merely indicates that for motor function the pathway is from some undisclosed place in the anatomical space of the brain through 'this or that nerve' to the muscles. Nevertheless, it seems plausible to assume that he conceived the human *sensorium* as the seat for both functions, the first of which (sensation) is a passion because initiated by an external, material cause (e.g., rays of light), the second of which (locomotion) is an action because initiated by an internal, efficient cause (soul as agent, not instrument).

Fortunately, this plausible answer finds support in Newton's suppressed 'Conclusio', a draft intended to be included at the end of the the second, 1713 edition of *Principia mathematica*.[9] Part of that draft includes words to the effect that motions by which the coat of the retina is agitated by rays of light may be propagated to the *sensorium* (add: through the optic nerves which are sensory nerves) and that similar motions may be propagated from the *sensorium* (add: through the motor nerves) to a 'substance' (i.e., motive power) in the muscles that dilates and, hence, contracts (i.e., swells) the muscles.

(Question 2) Where in the anatomical space of the brain is the place of the human *sensorium*?

Unfortunately, Newton's texts are silent concerning the *sensorium's* specific location in the brain and, hence, no answer can be given to Question 2 from the texts he intended for a public audience. Nevertheless, during this inquiry some evidence has been adduced that provides data for a plausible answer. Recall, first, the statement in Newton's 'Of Colours' in which he mentioned 'that precious organ', which is in 'ye center' or 'midst' of the brain, where 'the substance here is most pure'.[10] Recall, second, that Briggs, in his 1683 tract, wrote that the images^{b} of '*colour'd objects*' are transmitted to the '*Meditullium* of the brain (or common sensory)'.[11] Recall, finally, that in *Cerebri anatome* Willis localised the function of internal sense in the brain's two substances: the grey (cortical) substance as the seat of memory and the white (medullary) substance as the seat of common sense and imagination.

To the latter (white) substance, 'Species of sensible things' are carried from the external senses to the common sensory, namely, the corpora striata in '*the middle of the Brain*'[12] and there represented on the white wall or coat of the corpus callosum. And in the interior space of that anatomical substrate, which is the seat of imagination, those 'Species' (images^{b}) are beheld by phantasie and intellect, the knowing

[9] See *supra* Pt.I.1.2, comment to Text X.

[10] See the full quotation *supra* Pt.III.3.2, p. 80 and n.106, which provides evidence that by *c*.1666 Newton had read Willis's *Cerebri anatome*.

[11] See *supra* Pt.III.3.1, p. 71.

[12] Willis, *The Anatomy of the Brain* [1664 tr. 1681], p. 102 (italics mine). Note also that the phrase 'middle of the brain' may denote the brain's 'middle marrow' or white substance (see pp. 93 and 130) and therefore refer only obliquely to the two containers or mechanisms, the corpora striata and corpus callosum.

powers of the two souls (*anima* and *animus*). Since Willis treated the middle of the brain as the place of the common sensory (taken in his extended sense) and since Briggs regarded the '*Meditullium*' or middle of the brain as equivalent to the common sensory, it seems plausible to conclude that Newton's 'precious organ' in the 'center' or 'midst' of the brain is none other than what he called 'the *sensorium*'.

Although the answers given above to the four questions have different degrees of certainty, yet some additional support for the last two answers will emerge in the section that follows, As this part of the inquiry proceeds, it will be well to recall what was mentioned in the conclusion of Part II, namely, that Newton's method of physiological explanation involves distinguishing between effects (i.e., products or work), actions (i.e., micro-mechanical motions productive of effects) and unknown causes as disposing principles, powers or properties of the effects.[13] For, as will soon be evident, God is Newton's ultimate, as well as supreme disposing principle.

4.2 The Divine Spectator and the Cosmic Spectacle

In a number of his texts, Newton offers brief remarks about divine perception, which, he asserts, is direct and not like human perception, indirect.[14] Nevertheless, he describes both human and divine perception in visual terms: for the former, seeing, beholding and surveying; for the latter, seeing and observing. But since he also asserts that God is 'void of Organs, Members or Parts',[15] what does it mean to describe divine perception in visual terms? The answer is that God possesses *a disposing power* conceived not as a bodily eye but as an all-seeing intellectual eye. And this answer is suggested by two ways in which Newton applies to God the terms 'intelligent' and 'intelligence'.[16] For when he uses the first term to describe one of the attributes of God, he means that God *is* a rational being. And when he asserts that the uniformity in the solar and nervous systems is the result of 'intelligence and choice' or 'intelligence and wisdom',[17] he means that such uniformity *is the effect of* God's powers of intellect ('intelligence') and sound judgment ('wisdom') in the choice of means and ends.

However, to understand the relationship between intellect and choice, it is necessary, first, to discover the nature of Newton's theism, for, as James Collins has shown, there are many varieties of theism that develop as a response to different historical contexts and situations.[18] So what kind of theist is Newton? In terms of the

---

[13] See *supra* Pt.II.2.2, pp. 50–1.

[14] See *supra* Pt.I.1.2, Text VIII, comment to Text IX, Text XI and Text XII.

[15] See *supra* Pt.I.1.2, Text XII.

[16] See *supra* Pt.I.1.2, Text VIII, p. 21 and n106; Text IX, p. 22; and Text XI, p. 25 and n.137.

[17] See *supra* Pt.I.1.2, Text IX, pp. 21–2 n.101.

[18] See Collins, *God in Modern Philosophy*, who assayed the varieties of theism from Nicolaus Cusanus to the mid-twentieth century.

writings made public during his lifetime, an answer to this question may be found in his several attempts to infer the sufficient reason of the world from its rationality, as well as from its adaptation of means to ends. For example, in the first, 1687 edition of *Principia mathematica*, he inferred from the latter (adaptation of means to ends) that God positioned planets at different distances from the sun so that they would receive heat from it according to the proportion of their densities.[19] In the second, 1713 edition, this passage was excised; and in its place Newton added the concluding 'General Scholium', in which there is a brief inference from the rationality in the solar system, as well as from its adaptation of means to ends.[20] And in the 1706 *Optice* and the various editions of *Opticks (2)*, he included more extensive inferences, from which some portions have been extracted in Part I as Texts VIII, IX, XI and XII.

All these inferences constitute what usually is termed either the 'teleological argument' or 'argument from design'. In these arguments the term 'design' has a dual meaning, on the one hand, as *model* (means/plan) that has been conceived *in the mind of* God and, on the other hand, as his particular *intention* (ends/purpose) for that model. However, different versions of the argument from design depend on the kind of evidence used in the inferences. In contrast to the common argument that drew on evidence from artificial contrivances (e.g., a clock), Newton draws on evidence from natural contrivances, such as those described by himself or by some other unnamed natural philosophers and anatomists.[21] Hence, he may be classed as a naturalistic theist, whose inference scheme moves from natural contrivances to a supernatural contriver called 'God'.[22]

In the texts extracted from his teleological arguments, Newton describes God's substance as 'incorporeal' and his attributes as ever-living (eternally living), ever-present (omnipresent), powerful (omnipotent) and, as remarked above, intelligent (rational).[23] Although we can have no idea of God's substance, yet by investigating his plans and purposes, that is to say, his 'wise and excellent contrivances of things

[19] See Newton, *Principia mathematica* [1687], Bk. III, Prop. 8, in Cor. 5, p. 415. For Samuel Clarke's paraphrase of this argument, see Kassler, *Seeking Truth*, p. 182. See also Newton, 'The System of the World' [1728], *Principia mathematica (C-M)*, p. 566.

[20] See Newton, *Principia mathematica (C-M)*, Bk. III, General Scholium, pp. 543–4, and Newton, *The Principia*, Bk. III, General Scholium, pp. 585–6.

[21] For the historical background to, and development of this version of the argument, which was later termed the 'physico-theological method', see McAdoo, *The Spirit of Anglicanism*, pp. 240–315, *et passim*.

[22] According to Collins, *God in Modern Philosophy*, p. 54, this variety of theism, which he termed 'scientific', was 'the modern phenomenalistic substitute for two traditional arguments for the existence of God', namely, the *a priori*, ontological argument which moves from principles, reasons or causes to their consequences or effects, and the *a posteriori*, cosmological argument which moves from sense-grounded propositions in order to prove an initial premiss that a being exists and is the sufficient reason of the cosmos.

[23] See *supra* Pt.I.1.2, Text VIII, comment to Text IX, Text XI and Text XII. Note the absence of the attribute, all-knowing (omniscient), which is present, however, in Newton, *Principia mathematica (C-M)*, Bk. III, General Scholium, p. 545.

and final causes',[24] we may know indirectly that God exerts his prescient and beneficial power of intellect *to causally determine* his choice of means (plans) and ends (purposes) in regard to his acting either directly,[25] or indirectly by mediators[26] and by his ordained laws.[27] In all three ways, therefore, God remains the ultimate, though unknown first and final cause or sufficient reason of the material world—the cosmos.[28]

Consequently, when, in one text, Newton refers to the incorporeal God as an 'Agent',[29] this is to be understood in terms of agent-causation. And when Newton asserts that our soul, *animus*, is the image of God in us,[30] this, too, is to be understood in terms of the same kind of causation.[31] The basic assumption of this kind of causation is that willing is the cognitive function of judging that one object of consideration is to be set above others. Accordingly, this assumption equates volition

---

[24] See ibid., p. 546; cf. Newton, *The Principia*, Bk. III, General Scholium, p. 588: 'We know him ... by the wisest and best construction of things and their final causes'. Note that in the latter the translators seem not to have attended to terminology in the several arguments from design published during Newton's lifetime, for these do not use the term 'construction' either in the English or the Latin versions.

[25] See, e.g., *supra* Pt.I.1.2, Text XII, where Newton states that God is able to move bodies within his *sensorium* and thereby to form and reform parts in the solar system, including parts in the nervous system. According to Harrison, 'Newtonian Science, Miracles and the Laws of Nature', p. 549, Newton understood *direct* divine activity as interventions 'from our limited perspective only' and, as a consequence, for him, miracles have only 'epistemological, and not ontological status'. In short, God's interventions are unknown *efficient* causes.

[26] See, e.g., *supra* Pt.I.1.2, Text II, pp. 6 n.22, 7 n.31, 8 n.33 and Text IV, p. 13 n.63, where Newton identifies mediators as active principles or natural agents, that is to say, unknown *material* causes that produce effects such as cohesion, gravity, fermentation; see also Newton, *Opticks (C-R)*, Qu. 31, pp. 390, 394, 399. Consequently, such effects imply a rational contriver.

[27] In terms of the laws that can be discovered and demonstrated, the prime instance would be Newton's law of universal gravitation. But Newton attempted to discover other laws, including those which God has 'imposed on us' for moving our bodies; see *supra* Pt.II.2.1, p. 38 and see also n.34 below.

[28] OED 1: the world as an ordered and harmonious system (1690–1874).

[29] See *supra* Pt.I.1.2, Text XII.

[30] See *supra* Pt.I.1.2, Text IX and comment. For an earlier statement, see Newton, 'De gravitatione', p. 141: 'created mind (since it is the image of God) is of a far more noble nature than body' and 'the analogy between the Divine faculties and our own is greater than has been perceived formerly'. For the Platonic privileging of mind over matter, see the passage quoted from Cudworth *supra* Introduction, p. xx; see also Locke, *An Essay concerning Human Understanding*, p. 557: 'the whole intellectual World ... [is a] more beautiful World than the material [World]'.

[31] The scriptural passages in Genesis 1:26 and 27 state that 'God created man in his own image' but do not specify in what way; hence, the *imago Dei* was differently interpreted. For some interpretations in Newton's day, see Dockrill, 'The Heritage of Patristic Platonism', pp. 60–1, and Schneewind, *The Invention of Autonomy*, pp. 184–9, 203, 230. According to the version of Samuel Clarke, Newton's occasional mouthpiece, God made '*intelligent* Creatures ... so far like unto himself, as to endue them with those excellent Faculties of *Reason* and *Will*, whereby they are inabled to *distinguish* Good from Evil, and to chuse the one and refuse the other'; see Clarke, *A Discourse*, p. 62.

with intellectual preference and, hence, intellectualises volition.[32] For example, Locke defined volition as 'an Act of the Mind knowingly exerting the Dominion it takes it self to have over any part of the Man, by imploying it [volition] in, or withholding [*sic*] it from any particular Action'[33]; and therefore, the 'will', a 'Faculty', is determined by an antecedent intention to do or not to do.

To understand Newton's version of agent-causation, it is important to keep in mind his method of physiological explanation that derives from Galenic tradition. According to this method, agent-causation begins with a disposing principle which exercises its disposing power to causally necessitate its choice of plans and purposes and the voluntary actions that follow from them. The agent is *animus*, the power exerted is intellect,[34] and the choices necessitated are intentions called volitions that command 'at will' the doing or not doing an act. Note that it is not the will but the volitions—intellectual preferences—that command. The implication is that *animus*, the incorporeal disposing principle, is the person, that is to say, intelligence proper of a thinking being. And indeed, Newton himself states this, when he writes:

> Every soul that has perception is, though in different times and in different organs of sense and motion, still *the same indivisible person*. There are given successive parts in duration, coexistent parts in place, but neither the one nor the other *in the person of a man, or his thinking principle*; and much less can they be found in the thinking substance of God'.[35]

During this inquiry we have learned that corporeal substances also possess disposing principles, powers and properties. For example, in the sole text that provides insight into Newton's agent-causation concept in regard to muscle action, he writes: 'if there be any power in man to condense and dilate at will the æther, that pervades the muscle, [then] that condensation or dilation must vary the compression of the muscle, made by the ambient æther, and cause it to swell or shrink accordingly'.[36] Here, Newton's reference is to a motive power that swells the muscle on contraction. As indicated previously, this is a power that belongs to his corporeal disposing principle (*anima*), which is the unknown material cause of animal motion.

However, in the case of the incorporeal God and the incorporeal human soul (*animus*), it is their power of intellect that mediates 'the will', understood as an agent's volitions in regard to action. Consequently, there is an identity between

---

[32] See Bourke, *Will in Western Thought*, pp. 29 and 37, who also observed that with Locke, 'the term "preference" comes into continued use, in speaking of a willed choice'. For an instance of this, see *supra* Pt.III.3.3, p. 95.

[33] See Locke, *An Essay concerning Human Understanding*, p. 141.

[34] For a brief reference to the disposing power of divine and human intellect, see Newton, 'De gravitatione', p. 142, where he writes that intellect is infinite in God and 'in a finite degree belongs to us also; particularly since we do not move our bodies by our own independent power but through laws [of motion] imposed on us by God'.

[35] Newton, *Principia mathematica (C-M)*, Bk. III, General Scholium, p. 545 (italics mine). Cf. this version with Newton, *The Principia*, Bk. III, General Scholium, p. 587: 'Every sentient soul, at different times and in different organs of senses and motions [*sic*], is the same indivisible person. There are parts that are successive in duration and coexistent in space, but neither of these exist in the person of man or in his thinking principle, and much less in the thinking substance of God.'

[36] See *supra* Pt.I.1.2, Text II; Pt.II.2.1, pp. 38–40; Pt.II.2.2, p. 48; and for his revised account, Pt II.2.1, p. 44 n.84.

intellect and will, and, hence, Newton's stance may be described as intellectualist.[37] Like theism, however, there are a number of varieties of intellectualism that develop in response to different historical contexts and situations. The clue to Newton's variety is found in his conception of agent-causation, namely, that voluntary actions are the effect of antecedent intentions (volitions). And this conception points to a variety of intellectualism that Gilbert Ryle named 'the para-mechanical theory of the mind', because its causal proposition is that 'the bodily act' is 'the effect of a mental act of willing' to act.[38]

The conception of agent-causation described above also involves a patient; and the conception of patient will differ, depending on whether the agent (*animus*) is the incorporeal soul in humans or whether it is the incorporeal God. Since in humans the soul is embodied, the term 'patient' is to be understood as the corporeal *anima*, which is the instrument of the agent that senses and reflects, perceives and thinks, as well as the mediator between the agent that exerts the power and the body that moves. But since the incorporeal God is unembodied, the term 'patient' is to be understood as the spectacle that God beholds directly, namely, cosmic bodies that he moves directly or by mediators and by his ordained laws.

If the foregoing interpretation of Newton's conception of agent-causation is correct, then we may infer that his supreme disposing principle, God, is a functional principle. And in this he was not alone, for the conception of God as a functional deity was held by Locke and some other English empiricists, who repudiated Descartes's rationalist conception of God as a guarantor of knowledge and, thereby replaced his epistemological principle with their functional principle.[39] Yet neither Newton nor the other empiricists seem to have recognised the danger inherent in their own principle, for a functional deity is 'a highly expendable commodity once its functions can be explained by some other principles'.[40]

In addition to his functional principle, Newton's teleological arguments include two other principles, one, theological, the other, empirical. His theological principle, which is one of Newton's tenets of faith, is implied in the previous discussion of Newton's naturalistic variety of theism, namely, that there is a final cause or superintending providence ruling over events. But his empirical principle deserves comment here, because it is important for understanding part of his broader epistemological project. To begin with, note the particular passage in two of his arguments from design, where Newton asks (rhetorically): 'is it not evident from the

[37] As noted *supra* Introduction, pp. xi–xiii, this interpretation is at odds with the general consensus of commentators, who have associated Newton with voluntarism.

[38] See Ryle, *The Concept of Mind*, pp. 62–9, p. 64. For an interpretation of God's will as a thing-in-itself, rather than as a volition or antecedent intention determined by intellect, see Tamny, 'Newton Creation, and Perception, pp. 50–3.

[39] The nature of Newton's functional principle was missed by Burtt, perhaps because he used the term 'duties' not 'functions' when treating 'God's Present Duties in the Cosmic Economy', Burtt, *The Metaphysical Foundations of Modern Science*, pp. 291–7.

[40] For Locke, see Collins, *God in Modern Philosophy*, ('Empiricism and the Neutralizing of God'), pp. 97–107.

phenomena' or 'does it not appear from the Phænomena' that God exists.[41] What this passage signals is that Newton's inference scheme begins with *effects* derived not from phenomena themselves but from generalisations inferred from them and 'afterwards made general by induction'.[42] And hence, he arrives at his empirical principle, namely, that insofar as experience extends, there is rationality in the material world.

In two of his texts he expresses this principle as an ancient maxim: 'Nature doth nothing in vain'.[43] He also repeats this maxim in the second, 1713 edition of *Principia mathematica*, where it forms the first of four 'Rules' of reasoning in natural philosophy: 'Nature does nothing in vain, and more causes are in vain when less will suffice. For Nature is simple'.[44] Recall, however, that in his 1685 letter to Briggs, Newton made clear the double import of the maxim, 'Nature does nothing in vain', by glossing it as 'God ... has done nothing without reason'.[45] Since the maxim as glossed seems to represent a version of the principle of sufficient reason, Rule I states two principles, the second of which, 'Nature is simple', is Newton's principle of economy or simplicity.[46]

In both forms of expression, Newton's empirical principle—that there is rationality in the material world—is the result of his method of reasoning from experience. This method has been given a concise but important summary by Maurice Mandelbaum, according to whom reasoning from experience 'must seek the universal characteristics in the phenomena with which it deals', so that 'any characteristics common to all phenomena which we have experienced may be predicated of all objects whatsoever, whether or not these have been experienced, and whether or not they are experienceable'.[47] It would seem, therefore, that Newton's method includes two incompatible features, for

---

[41] See *supra* Pt.I.1.2, Text VIII, and Text XI, p. 24 n.133.

[42] See Newton, *The Principia*, Bk. III, General Scholium, p. 589, and Newton, *Ptincipia mathematica (C-M)*, Bk. III, General Scholium, p. 547 ('rendered general by induction').

[43] See *supra* Pt.I.1.2, Text VIII p. 20 n.103 and Text XI p. 24 n.133.

[44] See Newton, *The Principia*, Bk. III, Rule 1, p. 440; cf. Newton, *Principia mathematica (C-M)*, Bk. III, Rule 1, p. 398. Note that Newton's four Rules, which expound methodological principles, have been differently interpreted by commentators; see, e.g., Buchdahl, *Metaphysics and the Philosophy of Science*, p. 337, Holton, 'Presupposition in the Construction of Theories', pp. 89–93 and Shapiro, *Fits, Passions and Paroxysms*, pp. 49–50. Although Newton's formulation is considered 'classic', the principles concerned with economy, uniformity and evidence were expressed long before Newton; see, e.g., Crombie, *Robert Grosseteste*, p. 317.

[45] See *supra* Pt.III.3.1, p. 70.

[46] The principle of economy does not have the same meaning as the principle of parsimony (Ockham's razor), for this was a leading principle of nominalism, namely, that things not known to exist should not, unless absolutely necessary, be postulated as existing. Newton, however, cannot be classed as a nominalist, even though he did postulate a number of unobservables as real existents. For a brief discussion of this issue in relation to force, see Dijksterhuis, *The Mechanization of the World Picture*, pp. 480–3.

[47] Mandelbaum, *Philosophy, Science, and Sense Perception*, p. 81. See also Buchdahl, *Metaphysics and the Philosophy of Science*, p. 337, and Shapiro, *Fits, Passions, and Paroxysms*, pp. 40–8.

> On the one hand it endorses the view that all knowledge must start from sense experience and must proceed by forming generalizations which are based upon the repeated traits found within experience. On the other hand, it [also endorses the view that] it is legitimate to use these generalizations not only when dealing with phenomena which are in principle confirmable within experience, but that *it is also proper to use them inductively at* [sic] *generalizations concerning what transcends entities which are experienceable.*[48]

But this incompatibility disappears when Mandelbaum rephrased the words italicised in the above quotation as follows: 'whatever we find to be universally characteristic of that which our senses reveal we can generalize and hold to be true even beyond the limits of sense'.[49] To denote this move from observables to unobservables, he adopted the term 'transdiction'[50] instead of Newton's term, 'analogy of nature'.[51]

Commentators have tended to focus on Newton's move from observables to unobservables that are so minute as to be beyond the limits of sense. Indeed, Alan Shapiro defined transduction (his term for transdiction) as 'inference about unobservable microscopic components of bodies from knowledge of the observed laws and properties of macroscopic bodies'.[52] Yet, as Gerd Buchdahl pointed out, the transdiction process, as conceived by Newton, was intended to cover '*both the step "in depth"* from the realm of the observable to that of the unobservable (insensible particles: a kind of "analogical inference"), *and the step "in breadth"* from local regions of space to all regions'.[53]

In Part II of this inquiry the step '*in breadth*' was taken by Newton in his analogical inference from the functional anatomy of the retina in the interior space of the eye to the functional anatomy of the human *sensorium* in the anatomical space of the brain.[54] In this section evidence has been adduced that Newton also extended his generalisations concerning human sensory-motor functions to God as a divine spectator of, and causal agent in the cosmos. And, as will appear shortly, he further extended the step '*in breadth*' in order to provide what we might call a 'functional anatomy' of the divine *sensorium*.

Newton seems to have justified his generalisations concerning the deity, who transcends human experience, by claiming that 'all discourse about God is derived through a certain similitude from things human, which while not perfect is

---

[48] Ibid. (italics mine).

[49] Ibid., p. 87. For an instance of this, see *supra* Pt.I.1.2, Text V and comment.

[50] Ibid., pp. 61–2. For the 'first *explicit* recognition of the possibility of conjecturing the unknown from known phenomena', see Lloyd, *Polarity and Analogy*, pp. 337–41, who attributed this recognition to a dictum, formulated by Anaxagoras of Clazomenæ, which was approved and applied to unobservable atoms by Democritus of Abdera. For an interpretation of the dictum as preserved by Sextus Empiricus, see Kirk, Raven and Schofield, *The Presocratic Philosophers*, pp. 383–4.

[51] See *supra* Pt.I.1.2, Text II; see also Newton, *Principia mathematica (C-M)*, Bk. III, Rule 3, p. 398, and Newton, *The Principia*, Bk. III, Rule 3, pp. 441–2, p. 441.

[52] See Shapiro, *Fits, Passions, and Paroxysms*, pp. 4–5.

[53] See Buchdahl, *Metaphysics and the Philosophy of Science*, p. 337 (italics mine).

[54] A careful reader of Newton's texts, *supra* Pt.I.1.2, will notice that they contain many overt analogical inferences, sometimes foreshadowed by a phrase such as 'in like manner'.

nevertheless a similitude of some kind'.[55] It is important to note, therefore, that in the analogy of nature that leads to the divine *sensorium*, not all generalisations *within* the limits of sense are suitable for use *beyond* the limits of sense. As a consequence, during the move from the observable to the unobservable, certain features will be abstracted from that which has been found to be universally characteristic of phenomena within the limits of sense, as may be briefly instanced by imagining the steps '*in breadth*' that Newton might have taken in his analogy of nature between the retina in the interior space of the eye and the human *sensorium* in the anatomical space of the brain.

On the one hand, from reading the anatomical writings of both Willis and Briggs, as well as from experience as a witness during Briggs's dissection of the eye, Newton would have known that the exterior surfaces of the right and left eye, as well as the exterior surfaces of the right and left sides of the *sensorium* are lenticular, that is to say, they are shaped in the form of a lens. In addition, he would have known (as he clearly does in Text VII) that each of the two optic nerves 'meet' or 'unite', thus forming an X-shaped structure before they enter the brain, after which one nerve continues to the right, and the other to the left side of the corpora striata, the anatomical substrate that Briggs included in an illustration and that Willis identified as the common sensory.[56]

On the other hand, from the experimental demonstrations of Kepler, Descartes and others relating to the functional anatomy of the eye as the interior space of a *camera obscura*, Newton would have known (as he clearly does in Text III) that in the interior space of each eye, the coat of the retina is a receptor on which images[a] are formed. In generalising beyond the limits of sense, he could hold as true that in the middle of the brain, the coat of the *sensorium* is also a receptor on which images[b] are formed. And by this method he could infer that there is both a structural and a functional analogy between the retina and the *sensorium*, with this difference, that the light-sensitive substance in the retinal pores is adapted to receive stimuli from one external sense only, namely, the sense of sight, whereas the sensitive substance in the pores of the *sensorium* is adapted to receive stimuli from all the external senses.

In the steps 'in breadth' here imagined, certain features within the interior space of the eye itself would have to be abstracted, especially if they had not been universally demonstrated, such, for example, as whether the eye possesses one or more lenses in its interior.[57] Imagine, then, how much else would have to be abstracted when the final unobservable is the incorporeal God and his *sensorium*. Indeed, by necessity, only two features of the human *sensorium* would remain, namely, a place for the spectacle and a place beyond it for the divine spectator.

---

[55] See Newton, *Principia mathematica (C-M)*, Bk. III, General Scholium, p. 546, which is retained in Newton, *The Principia*, Bk. III, General Scholium, p. 589.

[56] See *supra* Pt.III.3.1, Fig. III.1, and Pt.III.3.2, Fig. III.2 and III.4.

[57] Newton retained the Keplerian model of the interior of the eye as a *camera obscura* with one lens (the crystalline humour), whereas Briggs and Willis added two more lenses. See *supra* Pt.I.1.2, Text III; Pt.II.2.1, pp. 50–1; Pt.III.3.1, p. 63; and Pt.III.3.2, p. 79.

Bearing this in mind, let us consider the texts in which Newton mentions the divine *sensorium*, beginning with Texts VIII and XI, in which he uses the term '*sensorium*' figuratively for God's omnipresence in infinite space 'as if [in] his sensory' or 'as it were in his Sensory'. This figurative usage suggests that Newton considers infinite space as the place of God, not as a property of divine incorporeal substance. But in the cancelled passage quoted in the comment to Text XI, he drops the similitude when he asks rhetorically:

> Is not the whole of Space the Sensorium of a Being Incorporeal, living, and intelligent, in that he sees distinctly and closely comprehends [i.e., takes in, contains] the most inward things themselves, and observes them wholly and thoroughly by their being present in him....

By identifying infinite space as the divine *sensorium*, Newton seems to imply that that space is a property of God's incorporeal substance.[58] Perhaps, realising this implication, he ordered the printer to cancel the page (or signature) containing the above text and to replace it with the figurative version in Text XI. However, later on, in Texts IX and XII, Newton drops the reference to space altogether and describes the properties of the divine *sensorium* not only as 'infinite' and 'boundless', but also as 'uniform'.[59] What, then, is Newton's position on this matter?

To answer this question, it is necessary first to understand what Newton means by space, an ontological problem that for many years has occupied the attention of numerous commentators with respect to his mathematical concept of space as something 'absolute'. According to Christopher Ray's useful analysis of the term,[60] Newton's concept has three distinct meanings: (1) space as an independent entity, that is, a real existent; (2) space as an entity possessing invariant properties; and (3) space as an irreducible element in a general account of motion and objects in space.[61] But is Newton's 'absolute' space synonymous with infinite space? This question is important, because some commentators, including Ray himself, write as if Newton

[58] In 1717 Samuel Clarke asserted that 'Space void of body, is the property of an incorporeal substance'; see Alexander, *The Leibniz-Clarke Correspondence*, p. 47. But according to Alexander, pp. xxviii–xxix, he seemed to have changed his mind, because in the French edition of the correspondence, published in 1720, the editor stated that Clarke had asked him to warn readers that 'when he speaks of infinite space or immensity ... he does not claim to take the term *quality* or *property* in the same sense as they are taken as those who discuss logic or metaphysic'; rather, 'he means only that space ... [is a mode] of existence of the Substance which is really necessary and substantially [i.e., in substance] omnipresent'. For Clarke's 1705 problematic conception of incorporeal substance as an unknowable substratum that underlies events, see Kassler, *Seeking Truth*, pp. 184–5.

[59] Although the terms 'infinite' and 'uniform' are also applied to space, yet there is a difference between the meanings of the term 'uniform' when applied to space and when applied to God, for which see *supra* Pt.I.1.2, Text XII, p. 25 nn.147 and 148.

[60] Ray, *Time, Space and Philosophy*, pp. 99–115.

[61] See ibid., p. 136.

supposed that the 'most beautiful system of the sun, planets and comets'[62] occupied the whole of infinite space.[63]

This supposition seems to have been made sometime before 1720, when Edmund Halley indicated that the 'System of the World, as it is now understood, is taken to occupy the whole *Abyss* of Space, and to be as such actually infinite', a doctrine that seemed to be confirmed by the 'Sphere of Fixt Stars'.[64] But in 'De gravitatione', a manuscript that is the product of mathematical lectures read at Cambridge University in the 1670s, Newton writes that the world ('mundam') is a 'limited' not an 'infinite being' and, consequently, *there are 'spaces beyond the world*, which we must suppose to exist (*since we imagine the world to be finite*), although they are neither revealed to us by God, nor known from the senses, nor does their existence depend upon that of *the spaces within the world*'.[65] Moreover, to demonstrate the dimensional claim that space 'extends infinitely in all directions', Newton resorts to the mathematics of asymptotic directions on a surface at a planar point, where all directions are asymptotic directions. Thus,

> ...if a triangle is actually drawn, its sides are always, in fact, directed towards some common point, where both would meet if produced, and therefore there is always such an actual point where the produced sides would meet, *although it may be imagined to fall outside the limits of the physical universe*.[66]

Then, in 1713, to identify the space occupied by the material cosmos, Newton invoked the fixed stars as an immovable reference frame at rest relative to absolute space,[67] whereas his laws of motion were supposed to hold in other reference frames 'with respect to the fixed stars'.[68] Accordingly, I conclude that Newton conceived cosmic

---

[62] See Newton, *Principia mathematica (C-M)*, Bk. III, General Scholium, p. 544. The translators of Newton, *The Principia*, Bk. III, General Scholium, p. 586, substituted 'elegant' for 'beautiful', though the former epithet is less congruous with Newton's own arguments from design—see, e.g., *supra* Pt.I.1.2, Text XI ('Order and Beauty')—or in some other such arguments by contemporaries—see, e.g., Kassler, *Seeking Truth*, pp. 194, 196–8.

[63] For Newton's brief critique of this supposition, see pp. 116–7 below.

[64] Halley, 'Of the Infinity of the Sphere of Fix'd Stars', p. 22, who, in this and the article following it, goes on to question the assumption of the infinity of the sphere of fixed stars. Note that by the term 'sphere', Halley probably meant region and neither the long-lived principle of nested or crystalline spheres, nor Kepler's life-long belief that the fixed stars were arranged in a spherical shell. For the two latter meanings, see Van Helden, *Measuring the Universe*, p. 4, *et passim*, and p. 63.

[65] See Newton, 'De gravitatione', p. 138 (italics mine).

[66] Ibid., pp. 133–4 (italics mine).

[67] For two experiments devised by Newton to demonstrate how 'true accelerations may be distinguished from relative [ones], and hence, the framework of absolute space identified, at least approximately, with that of the fixed stars'; see Hesse, *Forces and Fields*, p. 138. For a complication, see Burtt, *The Metaphysical Foundations of Modern Science*, pp. 294–5. For Newton's estimate of the distance of the region of the fixed stars and the problem it posed for the traditional Christian time frame, see Van Helden, *Measuring the Universe*, pp. 158–9.

[68] See Newton, *The Principia*, Bk. III, pp. 446, 450, 465–6, 471, 496, 502, 505, 525, 551–3, 556–7, 561. See also Jammer, *Concepts of Space*, p. 100, who pointed out that for Newton, the fixed stars 'were really "fixed," that is, not moving in space'. According to Van Helden, *Measuring the Universe*, p. 9, Aristarchus of Samos, in attempting to measure the sizes and distances of the sun

space as *an immense region* or finite part of infinite space and, therefore, that all parts of space have the same invariant properties,[69] that is to say, they are three-dimensional (forward-back, left-right, up-down); immovable (eternally fixed); uniform (all their constituent parts are alike) and penetrable (entities may be placed in parts of space). Consequently, both cosmic and infinite space may be described as 'absolute' in the second meaning given above. Finally, since the cosmic space occupies only a finite region in infinite space, this part alone may be described as 'absolute' in the third meaning given above.

If these conclusions are correct, then the place of the cosmic spectacle is that region in infinite space in which all corporeal 'things' are 'contained and moved'; but divine substance, being incorporeal, 'suffers nothing from the motion of bodies; bodies find no resistance from the omnipresent God'.[70] It is thus plausible to assume that Newton generalised the place of the spectacle from his much earlier conception of the place of the human *sensorium* as the center of the brain,[71] a conception which seems to have been reinforced by Willis's and, later, Briggs's equation of the common sensory with the center ('*Meditullium*') of the brain.[72]

And if he did so generalise, then it also is plausible to assume that Newton generalised the space beyond the framework of the fixed stars from the brain's white matter that surrounds the white matter of the human *sensorium*, the only difference being that the brain is bounded by the skull, whereas the space beyond the material cosmos is an unbounded vacuum and, hence, empty of all *corporeal* substances.[73] Recalling, therefore, the probability that Newton used his reflecting telescope as a model of the human *sensorium*, I conclude, finally, that the unbounded vacuum beyond the material cosmos is the place of the divine spectator, whose all-seeing intellectual eye beholds directly, that is to say, looks down upon the moving spectacle in the region of cosmic space.

And recalling that intellect, the rational knowing power of *animus*, perceives, as well as thinks, and that the space beyond the material cosmos is a vacuum, then it

---

and moon, utilised the fixed stars as the immobile reference frame against which all motions were plotted. That Newton was aware of this, see the first two paragraphs in Newton, 'The System of the World', *Principia mathematica (C-M)*, p. 549.

[69] See Newton, *Principia mathematica (C-M)*, Scholium to the Definitions, pp. 6–12, and Newton, *The Principia*, Scholium to the Definitions, pp. 54–61. Until the advent of non-Euclidean geometries, the dimensionality of space was understood as Euclidean.

[70] See ibid., p. 545, and ibid., pp. 587–8. Note that in the second version, the translation has: 'God experiences nothing from the motion of bodies, the bodies feel no resistance from God's omnipresence'. The choice of the words 'experiences' and 'feel' are unfortunate for two reasons: first, God is not the object of an action, i.e., he does not suffer (only God incarnate suffers); and second, bodies feel only if they have the property of sensitivity.

[71] See *supra* Pt.III.3.2, p. 80 and n.106.

[72] See *supra* Pt.III.3.1, p. 71 and n.106; see also p. 103 above.

[73] According to Jammer, *Concepts of Space*, pp. 22–3, since the Stoics maintained the existence of a void *outside* the *material* cosmos, it was 'owing chiefly to' their controversies that 'the problem of space could no longer be considered as one simple question, but had to appear under the form of two different considerations: space and void'.

also is plausible to assume that Newton might have made two further generalisations, which, if he did, will provide insight into the statement, repeated in three of his texts, that the divine *sensorium* occupies the whole of space, that is to say, cosmic space, as well as the vacuum beyond it. The two generalisations in question are as follows. First, in that part of the vacuum, which commences where the sphere of the fixed stars ends, God perceives 'things' directly, because his all-seeing eye of intellect surrounds and also 'comprehends' (i.e., takes in, contains) the material cosmos (perhaps like the pupil in the physical eye is surrounded by the iris). Second, in the unending space beyond his intellectual eye, God's 'thinking substance' (see p. 107 above) forms intellectual plans (design as model) and intellectual purposes (design as intention).

The plausibility of all three assumptions follows from Newton's analogy of nature, a method of reasoning from experience that has been described above. According to this method, he starts from what has been confirmed from experience, namely, the functional anatomy of the interior space of the eye, conceived, after Kepler, as an optical instrument. And then, by induction from this generalisation, he moves by steps in breadth from a region in anatomical space called 'the human *sensorium*' to all regions in infinite space called 'the divine *sensorium*'. As indicated previously, each step requires removing certain features from, or making adjustments to the original model. But in moving from a model of the eye to the human and then to the divine *sensorium*, he generates analogically his complete *sensorium* concept as a theoretical construct. In short, and to use the words of William Stanley Jevons, 'generalisation passes insensibly into reasoning by analogy'.[74]

However, as should now be evident, there is a correlation between the second part of Newton's *sensorium* concept and his mathematical concept of space as absolute. So if my interpretation of requires further support, it is to be noted that in the mid-1660s, when Newton was still an undergraduate at Cambridge University, Isaac Barrow delivered three sets of Lucasian mathematical lectures, one of which, in the second set, included a critique of the concepts of space propounded by philosophers such as Aristotle and Epicurus and Descartes and Thomas Hobbes.[75] In the course of this lecture, he made a number of observations, including that 'Mathematicians do, or lawfully may, draw Hypotheses suited to their Ratiocinations'[76]; and that the 'World is not proved to be actually infinite from the Imagination, but only to be greater in Power than any finite, (*i.e.* determinately finite) Thing'.[77]

[74] See Jevons, *The Principles of Science,* p. 627, who also observed, pp. vii, 9, that the 'great rule of inference' affirms that '*sofar as there exists sameness, identity or likeness, what is true of one thing will be true of the other*'. For Jevons, pp. 581, 583, Newton's *Principia* is 'the true Novum Organum' of scientific procedure, whereas Newton's *Opticks* is the best model of experimental and theoretical investigation.

[75] For the lecture in question, see *infra* Appendix.

[76] Barrow, *The Usefulness of Mathematical Learning*, p. 163.

[77] Ibid., p. 168

As for the principal hypothesis of his lecture, he sought to defend, also by a geometrical demonstration, that 'the Reality of *Space*' is distinct from 'Magnitude', that is to say, quantity of matter. And this defence involved a consideration of the problem of 'the true Extent of the Universe', namely, whether it is finite or infinite. In presenting his solution to this problem, he offered several reasons for the infinity of space, beginning with the following reason:

> ...since Matter may be finite, and God is infinite in Essence, he must subsist [also] beyond the Bounds of Matter, otherwise, he would be enclosed within its Limits or some Way bounded, and therefore could not be infinite. Therefore something is beyond, *i.e.* some sort of *Space*. And if God do not exist beyond the Bounds of Matter, ... it would follow, that we could not apprehend or conceive God to be immense.[78]

About a decade later, Newton, in 'De gravitation', adopted Barrow's hypothesis that space is real and distinct from matter.[79] And, in the course of developing this hypothesis, he returned to the problem of the extent of the universe to argue, as already indicated above, that space must extend infinitely beyond the limits of the 'world' and, as a consequence, the space of the material cosmos—cosmic space—must be finite.

4.3 Afterword: Is Absolute Space a Container?

Although the foregoing section represents the formal terminus of this inquiry, I have added an afterward as an answer to the many commentators who have assumed that *the whole of* infinite space is the place of the Newtonian cosmos.[80] Indeed, one more recent commentator, Christopher Ray, relied on historical data to support his assumption that Newton conceived 'the [spatial] universe' as '*an infinite arena containing matter everywhere*'.[81] As evidence for this assumption, he quoted a passage from the first of four letters that Newton wrote to Richard Bentley in 1692.[82] He then followed this quotation with the statement that a 'postulation of a finite spatial *container* for matter' would not help Newton's realisation that 'a tiny cluster' of fixed stars 'would tend to collapse in upon itself as a result of the gravitational attractions between the stars'.[83]

---

[78] Ibid., p. 171.

[79] See *supra* Introduction, p. xv.

[80] See pp. 112–3 above.

[81] See Ray, *Time, Space and Philosophy*, p. 177 (italics mine). According to Newton, *Principia mathematica (C-M)*, Bk. III, General Scholium, p. 544: '...and lest the systems of the fixed stars should, by their gravity, fall on each other, he [God] hath placed those systems at immense distances from one another'. See also Newton, *The Principia*, Bk. III, General Scholium, p. 586.

[82] See Cohen and Schofield (eds.), *Isaac Newton's Papers*, 'Letter I', pp. 280–90, who also include the remaining letters, all in facsimile from their first publication in 1756. Bentley, who initiated the correspondence, wrote in regard to Sermon 7, which formed part of the first set of his Boyle Lectures read between 7 March and 5 December 1692.

[83] See Ray, *Time, Space and Philosophy*, p. 178 (italics mine).

In making the above statements, Ray completely disregarded the context in which the historical evidence for his assumption occurs, for Newton's first letter states that *if* 'all the Matter of the Universe' were to be 'evenly scattered throughout all the Heavens' and *if* there were only 'meer natural Causes', *then* the solar system would become unstable (in some ways that Newton describes) and would require the intervention of 'an intelligent Agent', that is to say, a rational agent who is 'skilled in Mechanicks and Geometry'.[84] Indeed, had Ray read the fourth letter to Bentley, he would have discovered that Newton rejects the 'Hypothesis of deriving the Frame of the World by mechanical Principles from Matter evenly spread through the Heavens', because that hypothesis is 'inconsistent with my System'.[85]

Moreover, Newton does not mention the fixed stars or their collapse in his letters to Bentley; but he does state conditionally that *if* Bentley had proved 'the Finiteness of the Universe', *then* 'it follows, that all Matter would fall down from the Outsides, and convene in the Middle'.[86] And *if* 'all the Matter were divided at first into several Systems, and every System by a divine Power constituted like ours', *then* 'would the Outside Systems descend towards the Middlemost; so that this Frame of Things could not always subsist without a divine Power to conserve it'.[87]

However, later on in *Opticks (2)*, Newton does mention the collapse of the fixed stars, when indicating that 'the main Business of natural Philosophy is 'not only to unfold the Mechanism of the World, but chiefly to resolve these and such like Questions', viz.:

> What is there in places almost empty of Matter, and whence is it that the Sun and Planets gravitate towards one another, without dense Matter between them? Whence is it that Nature doth nothing in vain; and whence arises all that Order and Beauty which we see in the World? To what end are Comets,[88] and whence is it that Planets move all one and the

---

[84] The quoted passages are from Cohen and Schofield (eds.), *Isaac Newton's Papers*, 'Letter I', pp. 281–2, 284, 287.

[85] Ibid., 'Letter IV', p. 310. Newton's description suggests that the 'Hypothesis' in question is that of Descartes.

[86] Ibid., 'Letter III', p. 305.

[87] Ibid., p. 306. By 'several Systems', Newton alludes to a plurality of worlds in infinite space, a conception that also would be incompatible with his celestial mechanics; for his stance on this matter, see Dick, *Plurality of Worlds*, pp. 143–7; see also p. 123 below. That conceptions of a plurality of worlds have had 'complete sway' up to the 'present day' (1920) was pointed out by Schlick, *Space and Time in Contemporary Physics*, pp. 67–75, who also provided a short demonstration of the incompatibility of Newton's system with conceptions of multiple systems.

[88] This question is answered by Newton, *Principia mathematica (C-M)*, BK. III, p. 530, and Newton, *The Principia*, Bk. III, p. 572: 'I suspect ... that it is chiefly from the comets that spirit comes, which is indeed the smallest but most subtle and useful part of our air, and so much required to sustain the life of all things with us.' The first clause may be original with Newton, but the remainder expresses a widespread view among certain physiologists who believed that some kind of chemical spirit in the air was necessary for life. But behind them all was the work of Robert Hooke, who in the early 1660s had been interested in respiratory physiology, an interest that led him to postulate in 1665 that a nitrous component in the air was an essential natural agent for combustion and respiration and in 1667 to collaborate with Richard Lower on cardiopulmonary experiments; see Lysaght, 'Hooke's Theory of Combustion' and Brown, 'Richard Lower', p. 525.

same way in Orbs concentrick, while Comets move all manner of ways in Orbs very excentrick; and *what hinders the fix'd Stars from falling upon one another?*[89]

In posing these questions Newton suggests that God, his ultimate disposing principle, functions not only to create the order and beauty of the solar system, but also to sustain that order by, e.g., preventing the fixed stars from collapsing in the middle of cosmic space.[90] And here, of course, lies one of the dangers in conceiving God as a functional principle, since over time, some, but not all of the questions Newton poses above will be answered by other investigators. Even so, Newton's questions, as well as his letters to Bentley, clearly point to the two main tenets of his faith, namely, that God is both first and final cause of the material cosmos. And for commentators to ignore these tenets is to misrepresent Newton's thought.

Now, Ray's assumption that the 'universe' is 'an infinite arena containing matter everywhere', implies that the 'arena'—infinite space—is a container. While admitting to being 'sympathetic up to a point with anyone who might say that the *Principia* does not justify attributing the "container" idea directly to Newton', Ray himself would 'justify the use of this idea in the Newtonian world-view, given his [Newton's] belief that God created the material world and placed it in space at a given time.'[91] Note, therefore, that numerous other commentators have used the terms 'container' or 'receptacle' to describe Newtonian absolute space.[92] In 1928, for example, J. H. Randall, Jr. wrote that God, by existing, constitutes absolute space as the container in which all motions take place and, consequently, this 'Divine Sensorium or Mind sustains the entire field of physics'.[93] And according to a more recent commentator, the container conception of dimensionality was 'canonized by Isaac Newton (1686), who held that space and time are infinite and independent of physical bodies which exist in them', the implication being that such containers contain the sensible contents of experience.[94]

---

[89] Newton, *Opticks (C-R)*, Qu. 28, pp. 362–70, p. 369 (italics mine).

[90] Regarding the latter function, see Burtt, *The Metaphysical Foundations*, p. 295, who pointed out that in *Principia mathematica* Newton had confined himself to observing that in order to prevent the collapse of the fixed stars, 'God had set these stars at immense distances from one another' (see p. 116 n.81 above), an expedient that would 'hardly suffice through all the ages of time'.

[91] Ray, *Time, Space and Philosophy*, p. 240 n.2, see also p. 99.

[92] See, e.g., Aronson, Harré and Way, *Realism Rescued*, p. 84, who described Newton's ontology as based on 'forces, atoms and a (notorious) container version of the absolutist view of space and time'. See also Buchdahl, *Metaphysics and the Philosophy of Science*, pp. 549 n.5, 582, 584, 587–9, 595 and n.3, 647, 666–7, for Emanuel Kant's opposition to the doctrine of absolute space 'under the image of the boundless receptacle' of 'possible things'.

[93] Randall, quoted in Strong, 'Newton and God', p. 148. He also claimed that God constitutes both absolute space and absolute time and that these are his *sensorium*. Although it is widely held that these two absolutes constitute the ultimate structure of Newton's mathematical universe, absolute time did not enter into the subject of this inquiry. According to Ray, *Time, Space and Philosophy*, pp. 102–3, Newton's argument for absolute space 'may be readily extended to give us a "Newtonian" argument for absolute time'. Although he did not provide that argument, his statement suggests that absolute time, for Newton, is space in one dimension. For a different suggestion, see *infra* Appendix, p. 151.

[94] See Hundert, *Philosophy, Psychiatry and Neuroscience*, p. 87.

But infinite space, by definition, is boundless,[95] that is to say, measureless, but not dimensionless; and, therefore, to describe *the whole of* absolute space as a container is a contradiction in terms.[96] Moreover, historical evidence suggests that caution should be exercised when applying the term 'container', since it could be used to describe Aristotle's influential conception of space and place as a physical or entitative, not a mathematical or ideal relationship,[97] a conception that was under scrutiny and criticism in Newton's day.[98] According to this conception, all individual determinations of place are to be united in '*one whole space*', so that 'the one, all-embracing space *contains* the particular places within it as constituent parts'. Accordingly, 'general space only signifies *the last most external shell beyond which there can be neither space nor bodies*'.[99] In other words, general space is 'the inner boundary of *the containing receptacle* ... which *itself is not contained in any further receptacle*'.[100] Indeed, there is no space outside that receptacle, because, for Aristotle, infinity denoted imperfection.[101]

Later on, however, theologians such as the third-century Platonist, Novatian, or the thirteenth-century Scholastic, Alexander of Neckham, thought otherwise. For both men conceived God as *an infinite container*. According to the former, God is

> ...immense and without limit, not one who is enclosed in a place but one who encloses every place, not one who is in a place, but rather in whom every place is, one containing everything and clasping it together, so that accordingly he neither ascends or descends since he himself contains and fills everything.[102]

According to the latter, God's 'immensity' is the spatial 'frame' in which everything is contained, so that 'everything' that God has created 'is conserved in this Self, in which [quoting St. Paul] we live and move and have our being'.[103] Later on, in his dialogue between God and the soul, Marcilio Ficino, the fifteenth-century leader of

---

[95] See *supra* Pt.I.1.2, Texts VIII, IX, XI, and XII.

[96] According to Van Helden, *Measuring the Universe*, p. 3, telescopic astronomy made it possible to subject to scientific scrutiny the traditional scheme of sizes and distances of bodies in the solar system. This scrutiny thus allowed the emergence of a new consensus based on direct measurements for the apparent sizes of the planets and the all-important distance between sun and earth; and it also helped make the solar system itself inconceivably large.

[97] For historical treatments of space in classical antiquity and in the period from the Middle Ages to the seventeenth century, Algra, *Concepts of Space*, and Grant, *Much Ado about Nothing*.

[98] For one critic, see *infra* Appendix.

[99] See Cassirer, *The Individual and the Cosmos*, p. 181 (italics mine). For Aristotle's cosmos, see also Crowther and Barker, 'Training the Intelligent Eye', pp.454–6.

[100] See Jammer, *Concepts of Space*, pp. 17 (italics mine).

[101] See Davenport, 'The Catholics, the Cathars, and the Concept of Infinity', p. 265 n.5.

[102] Quoted in Westfall, *Never at Rest*, p. 318.

[103] Quoted in Davenport, 'The Catholics, the Cathars, and the Concept of Infinity', p. 271, who also pointed out that by conceiving God as an infinite container, Neckham and other scholastic theologians were required to scrutinise the conception of divine immensity, which had been introduced at the Fourth Lateran Council to mean that God is without or beyond all spatial dimension.

the Florentine Academy in Italy,[104] repeated the conception of God as a container when he had God proclaim:

> I fill and penetrate and contain heaven and earth; I fill and am not filled because I am fullness itself. I penetrate and am not penetrated, because I am the power of penetration. I contain and yet am not contained, because I myself am the faculty of containing.[105]

Insofar as the above examples describe God as an infinite container of other substances, there would be the danger of pantheism, since the infinite container idea implies that God is omnipresent only substantially. But Ficino avoided this danger by adding that God is also present virtually, because he has a power ('faculty') of containing.[106]

And, finally, in 1665 Barrow, in one of his Lucasian lectures on mathematics, several times applied the term 'Capacity' to space,[107] first, to describe it as 'a certain Extension and indefinite Capacity'[108]; next to describe Aristotle's *'immoveable Vessel'* as a 'Receptacle of immense Capacity'[109]; then, in an attempt to answer the question, what is space?, he replied: '...*Space* is nothing else but the *mere Power, Capacity, Ponibility*, or (begging pardon for the Expressions), *Interponibility of Magnitude*'[110]; and in attempting to explain his meaning, he stated that space has a 'Capacity' or 'Power' to receive or contain 'no actual but only potential Figures, Dimensions and parts consentaneous to its Nature'.[111] And he afterwards added that '*Space* is not a particular, but a kind of general and indefinite Capacity'.[112]

This brief historical digression brings us back to Newton's claim, quoted in the previous section, that 'in' God 'are all things contained and moved',[113] a claim that modifies St. Paul's response to the Stoics and Epicureans in Acts 17.28: 'in him we live and move and have our being'.[114] How is Newton's modification to be

[104] See Kristeller, *Renaissance Thought II*, p. 95, who described Ficino's Platonism as 'a highly complex and unique system of thought that cannot be taken as a mere repetition of the thought of Plato or of the ancient Neoplatonists'.

[105] Quoted in Cassirer, *The Individual and the Cosmos*, pp. 190–1.

[106] See Newton, *Principia mathematica (C-M)*, General Scholium, Bk. III, p. 545: God 'is omnipresent not *virtually* only, but also *substantially*; for virtue [power] cannot subsist without substance'. For the scholastic tradition implied in this passage, see also Letvin, 'Newton and Scholastic Philosophy', pp. 70–5 (who focused on 'virtually', not 'substantially').

[107] See *infra* Appendix, pp. 146 and 150.

[108] Barrow, *The Usefulness of Mathematical Learning*, p. 164.

[109] Ibid., p. 165.

[110] Ibid., p. 176.

[111] Ibid., p. 177. For the meaning of this usage, see OED 1a obs.: Ability to receive or contain, holding power (1481–1702).

[112] Ibid., p. 182, which, perhaps, implies something like Plato's receptacle (see p. 37 below).

[113] See p. 114 and n.69 above.

[114] In the 1713 edition of *Principia mathematica*, Newton supported this statement by citations to one ancient source, St. Paul, and a few other sacred writers. But in the editions that followed, the first of several ancients cited is Pythagoras, whereas the first of the sacred writers is St. Paul. Newton indicates that his source for the opinion of Pythagoras was Cicero, *The Nature of the Gods*; but in Bk. I, p. 81 of that source, a reader will discover that Velleius the Epicurean, one of the

understood? If my interpretation of that the divine part of his *sensorium* concept is correct, then his claim refers only to cosmic space and not to the immense vacuum beyond that space. And if this is so, then, with appropriate qualification, one could describe the region of cosmic space as a container of the solar system, because that system is immovably 'framed' (Newton's term) *not by* divine incorporeal substance *but by* the corporeal substance of the fixed stars.

But Ray notwithstanding, God cannot be described as placing 'the material world' in infinite space. For, on the one hand (and to repeat), the material world occupies an immense but finite region in infinite space; and, on the other hand, as Newton tells us, God 'form'd' the absolutely solid particles (atoms) that bring the world into existence. Here are his words:

> God, in the Beginning, form'd Matter in solid,[115] massy, hard, impenetrable, moveable Particles, of such Sizes and Figures, and with such other properties, and in such Proportion to Space, as most conduced to the End[116] for which he form'd them; and that these primitive Particles being Solids, are incomparably harder than any porous Bodies compounded of them, even so very hard, as never to wear or break into pieces; no ordinary Power being able to divide what God himself made one in the first Creation. While the Particles continue entire, they may compose Bodies of one and the same Nature and Texture in all Ages. But should they wear away, or break in pieces, the Nature of Things depending on them, would be changed. ... And therefore, that Nature may be lasting, the Changes of corporeal Things are to be placed only in the various Separations and new Associations and Motions of these permanent Particles, compound Bodies being apt to break, not in the midst of solid Particles, but where those Particles are laid together, and only touch in a few Points.[117]

The above story is expressed in terms of Newton's mature thought concerning the origin of the world, as is evident by comparing it with an earlier 1675, more hesitant story that

> Perhaps the whole frame of nature may be nothing but various contextures of some certain ætherial spirits, or vapours, condensed as it were by precipitation, much after the manner, that vapours are condensed into water, or exhalations into grosser substances, though not so easily condensible; and after condensation wrought into various forms, at first by the immediate hand of the Creator; and ever since by the power of nature, which, by virtue of the

---

participants in Cicero's imaginary conversation, sets up in order to criticise the supposed opinion of Pythagoras 'that mind was present and active throughout the whole universe and that our own minds were a part of it'. This opinion, of course, is neither commensurate with that of St. Paul nor with the doxographical tradition concerning Pythagoras; see Kirk, Raven and Schofield, *The Pre-Socratic Philosophers*, pp. 214–38. Consequently, Newton's citations to Pythagoras and St. Paul do not seem to be an instance of what Paolo Cassini had identified as Newton's non-dogmatic principle of textual authority; see *supra* Introduction, p. x.

[115] OED *a* I.1: Free from empty spaces or interstices (c.1391–1881); i.e., as opposed to solids that are composites or dense bodies with pores.

[116] I.e., design as intention.

[117] Newton, *Opticks (C-R)*, Qu. 31, pp. 375–406, p. 400; see also, pp. 403–4. In the process of forming compound bodies, there is no real coalescence of atoms; rather, they simply come into contact with each other and always retain their own shape and individuality. This, too, was the opinion of the ancient atomists; see Kirk, Raven and Schofield, *The Presocratic Philosophers*, p. 426.

command, increase and multiply, became a complete imitator of the copies set her by the protoplast.[118] Thus perhaps may all things be originated from æther.[119]

Here, Newton accounts for changes in motion with reference to corporeal but unobservable 'ætherial spirits, or vapours'. And the extracts presented in Part I from the same textual source inform us that these unobservable spirits have dispositions called 'sociability and unsociability'.[120] Extracts from the same textual source also inform us that that when considering the action of animal motion, Newton postulates ætherial vital spirits in the circulatory system and ætherial animal spirits in the nervous system; and in 1675 these together constitute his disposing principle of life, sensation and motion (*anima*).[121]

Newton's mature story, quoted above, represents an emendation of an earlier version included in the 1706 *Optice* as Query 23.[122] Despite certain differences, both versions account for changes in motion with reference to unobservable primordial atoms and their dispositions of association and separation.[123] Note also that in drafts for Query 23, written in English about 1705, there is the following declaration:

Life & will are active Principles [i.e., natural agencies] by wch we move our bodies, & thence arise other laws of motion unknown to us. And since *all matter duly formed is attended with signes of life* & all things are framed wth perfect art & wisdom & *Nature does nothing in vain*; *if there be an universal life* & all space be *the sensorium* of a thinking being who by immediate presence perceives all things in it as that wch thinks in us perceives their pictures in the brain, [then] the laws of motion *arising from life or will* may be of universal extent.[124]

This declaration increases the possibility that Newton's two brief stories of the origin of the world owe a debt to the 'likely story' in the *Timaeus* of Plato.

According to this story, before the 'world' or 'cosmos' came into existence,[125] there were 'three distinct realities', being, space and becoming.[126] In the space of becoming a pre-existent 'receptacle' (*chora*), provides the possibility for the embodiment of ideas. According to Evgeny Zaitsev, the receptacle itself has 'almost' no materialistic connotations, because it was conceived by Plato 'not as a material substance from which things are produced', but as 'a kind of "general place" ...

[118] OED protoplast^2: The first former, fashioner or creator (1600–1872).

[119] See Newton, 'An Hypothesis explaining the Properties of Light', Cohen and Schofield (eds.), *Isaac Newton's Papers*, p. 180. Note the analogy to the water cycle.

[120] See *supra* Pt.I.1.2, Text II and comment.

[121] See *supra* Pt.I.1.2, Text II; see also *supra* Pt.II.2.1, p. 45.

[122] See Newton, *Optice*, Qu. 23, pp. 322–48, pp. 343–4. There is no comment on this emendation in Koyré, 'Les Queries de l'Optique'.

[123] I.e., attraction and repulsion; see *supra* Pt.II.2.2, p. 54.

[124] Quoted in Westfall, *Never at Rest*, pp. 647–8 n.59 (italics mine).

[125] See Lee (tr.), *Plato Timaeus*, p. 41: 'as for the world—call it that or cosmos or any other name', 'we must ask ... whether it has always existed and had no beginning, or whether it has come into existence and started from some beginning'; and the 'answer is that it has come into being; for it is visible, tangible, and corporeal, and therefore perceptible by the senses'.

[126] Ibid., p. 72.

where forms of things appear as images of eternal forms (ideas)'.[127] In this general place or receptacle,[128] the skilled workman or 'demiurge' reduces indeterminate, chaotically-moving matter to determinate, orderly moving matter, first, by constituting it into four basic elements or particles 'too small to be visible'[129] and, then, by giving these moveable particles the shape of four of the five regular geometrical solids. In effecting the latter, the demiurge introduces 'measureable relations, internal and external, among them, to the degree and extent that they were capable of proportion and measurement'.[130]

Having brought these unobservable geometrically-shaped particles to 'the exactest perfection permitted by the willing consent of necessity',[131] the demiurge initiates a copy process that will reproduce the 'plan' or 'pattern' (design as model) of the perfect 'eternal Living Being'.[132] Initially, this copy process is carried out by the combinations and transformations of the unobservable particles, which then form larger compounds and aggregates.[133] And thus, through the 'providence' or 'purpose' (design as intention) of the demiurge, the world comes into existence as 'a single visible living being, containing within itself all living beings of the same natural order'.[134] Indeed, this cosmos 'was and is and will continue to be' the 'only creation' of the 'eternal Living Being'.[135]

Newton, too, believed in only one creation,[136] and he was not alone in this. For although Kepler had derived the actual distance of the fixed stars (a distance that was indeed immense), yet he did not accept the position of Giordano Bruno who

> ...made the world so infinite that [there are] as many worlds as there are fixed stars. And he made this our region of the movable [planets or wandering stars] one of the innumerable worlds scarcely distinct from the others which surround it, so that to somebody on the Dog Star [the largest of the fixed stars] ... the world would appear from there just as the fixed stars appear to us from our world.[137]

---

[127] See Zaitzsev, 'The Meaning of Early Medieval Geometry', p. 541 and n.46. For a more accurate description of the 'forms' (*paradeigma*), see *infra* Conclusion p. 131.

[128] For the receptacle, see Lee (tr.), *Plato Timaeus*, pp. 67, 70–3, and the comparison, p. 69, between the receptacle and the 'nurse' or 'mother', i.e., the female matrix or womb.

[129] Ibid., p. 79.

[130] I.e., 'determinate cause', see ibid., p. 96. By introducing such relations, the demiurge makes the cosmos intelligible to the intellect, not merely perceptible by the senses (see p. 122 n.125. above).

[131] Lee (tr.), *Plato Timaeus*, pp. 79 and 66.

[132] Ibid., p. 51; see also pp. 43, 54–5.

[133] Ibid., p. 73. Afterwards the copy process is carried out by created divinities.

[134] Ibid., p. 43.

[135] Ibid., p. 43, see also p. 78.

[136] See p. 117 n.87 above.

[137] Quoted in Van Helden, *Measuring the Universe*, p. 63 (additions in square brackets mine). See also Schlick, *Space and Time in Contemporary Physics*, pp. 67–9.

Perhaps because Newton did not wish to derogate from God's power, he conditionally allows that the fixed stars might be centers of other solar systems and that if this were so, then God would be the governor of those systems as well as our own.[138]

Although Newton's two origin stories both contain features of Plato's story, two such features deserve notice. On the one hand, the earlier 1675 story suggests overtly that God 'wrought' pre-existing matter into various forms, whereas the later story merely implies this in the statement that God 'form'd' 'Matter' into atomic solids of different sizes and shapes and with dispositions to associate and separate.[139] On the other hand, despite differences between the two stories, both agree in conceiving the unobservables as unknown material causes, namely, natural agencies that are instinct with life and, hence, are sources of activity in nature. Nevertheless, both features of Newton's stories are problematic. For, just as Plato's perfect 'eternal living Being'is presented as a skilled worker (demiurge), so too Newton's God is presented as a skilled contriver who impresses form on pre-existent matter.[140] Absent from his story is a creator who *first* generates shapeless matter *and then* shapes it into various forms. We might ask, therefore, what happened to the first of Newton's two main tenets of faith, namely, God as the unknown first cause?

The answer must be that Newton's physico-theological versions of the argument from design start from the belief that the cosmos was created and generated from a plan (design as model) *in the mind of* God.[141] Accordingly, not only the process of creativity, but also the ideal plan itself cannot be known directly, because, as Newton's contemporary, Locke, proclaimed:

> If you do not understand the Operations of our own finite Mind, the thinking thing within you, do not deem it strange, that you cannot comprehend the Operations of the eternal infinite Mind, who made and governs all things, and whom the Heaven of Heavens cannot contain.[142]

Nevertheless, as Newton himself points out, we can understand God's purposes (design as intention) by investigating 'his most wise and excellent contrivances of

---

[138] See Newton, *Principia mathematica (C-M)*, Bk. III, General Scholium, p. 544, and Newton, *The Principia*, Bk. III, General Scholium, p. 586.

[139] See also Newton, *Opticks (C-R)*, Qu. 31, pp. 375–406, p. 401, where he notes that the active principles that cause physical phenomena such as gravity, cohesion and fermentation are 'general Laws of Nature, by which the Things themselves are form'd'.

[140] See *supra* Pt.I.1.2 Text VIII (fabricata/fashioned), Text IX (fabricatæ/framed, conformatio prima, first shaping), Text XI (contrived). For an exception, see *supra* Pt.I.1.2, comment to Text IV, p. 14 n.72.

[141] Without this assumption, Newton's teleological argument leads only to an intelligence which produces system, order and teleology in the cosmos, thereby leaving a dualism between supracosmic mind and the material to be shaped.

[142] See Locke, *An Essay concerning Human Understanding*, p. 630. Note that the last six words echo some scriptural texts, for which see *infra* Appendix, p. 143 and n.19.

things'.[143] But to achieve this understanding, it is necessary to attend to 'the main Business of natural Philosophy', namely, 'to argue from Phænomena without feigning Hypotheses, and to deduce Causes from Effects, till we come to the very first Cause, which certainly is not mechanical.[144] For 'though every true Step made in this Philosophy brings us not immediately to the Knowledge of the first Cause, yet it brings us nearer to it, and on that account is to be highly valued'.[145]

[143] Newton, *Principia mathematica (C-M)*, Bk. III, General Scholium, p. 546; cf. and Newton, *The Principles*, Bk. III, General Scholium, p. 588.

[144] Newton, *Opticks (C-R)*, Qu 28, pp. 362–70, p. 369.

[145] See *supra* Pt.I.1.2, Text XI.

Conclusion

In the foregoing inquiry I hope to have shown why an evidence-based interpretation of Newton's *sensorium* concept must follow the order of knowledge required by the method of reasoning from experience that he called 'the analogy of nature', otherwise there will continue to be guesses about, as well as misinterpretations concerning the meaning of his concept.[1] The interpretation of Gerd Buchdahl is an instance of the latter problem. For although he had grasped that Newton's analogy of nature includes regions of space and, hence, is not restricted to analogical inferences concerning micro-mechanical motions of unobservable corporeal substances,[2] yet he went astray when considering the question: How are we to understand the way God 'constitutes' space? And the answer he put into the mouth of Newton was: 'this can be made intelligible through the conception of God's *sensorium*, something which .... in turn [can] be grasped by analogy with ... the *sensorium* of man'.[3]

However, instead of elucidating that part of Newton's *sensorium* concept relating to humans, Buchdahl began with Locke's theory of perception, because, as indicated previously,[4] he believed that Newton owed a debt to that theory, which he summarised as follows:

> We are acquainted only with our 'ideas' (Newton calls them 'images'), which alone are *in or before our minds*. At the same time, we do see things in space. We may describe this by saying that the 'idea' represents things as 'appearing in space'. Space is thus an integral aspect of **the *sensorium*, which is, of course, nothing physical**, and is quite distinct from the sensory organs. In virtue of **the sensorium**, what is, metaphysically speaking, 'in the

[1] For some of these guesses, see *supra* Introduction, p. v nn.2 and 3.

[2] See *supra* Pt.IV.4.2, p. 110.

[3] Buchdahl, *Metaphysics and the Philosophy of Science*, p. 595. His answer assumes that readers contemporary with Newton would have had up-to-date knowledge about the human *sensorium*, but this is doubtful.

[4] See *supra* Pt.III.3.3, p. 93.

J. C. Kassler, *Newton's* Sensorium*: Anatomy of a Concept*, Archimedes 53,
https://doi.org/10.1007/978-3-319-72053-1_5

> mind', is described as 'appearing in space'; seeing things **in our *sensorium*** is to see things in space.[5]

He then provided a summary of Newton's theory of perception, which

> Evidently ... is this: 'Space' subsists in the context of 'seeing things in space' in the way in which 'ideas' *represent* things in space, out there, although there *is* no real space; it can only be ideal, since the 'ideas' themselves are not 'out there'. So *our* space is 'constituted by our 'seeing things out there'. However, Newton at the same time affirms that this is only a 'simile' for that one real space in which—as real things—everything subsists. As a result, he argues that just as everything subsists in one space *for us*, in virtue of and in the context of our perceptions, so there is one real space, in virtue of God's perceptions; God 'seeing the things themselves', rather than their ideas.[6]

On the basis of these two summaries, he concluded that because the human *sensorium* yields 'a kind of' ideal space, 'it is an analogy for God's *sensorium*', with this difference that God sees the things themselves directly, not indirectly.

The problem with Buchdahl's interpretation begins with his statement in the first summary (above) that the human *sensorium* is 'nothing physical'. And this suggests that, for him, the term denoted a functional space in which ideas (images[b]) subsist 'in the mind', thereby implying that mind and ideal space are co-extensive. But as has been learned during the course of this inquiry, Newton conceives the human *sensorium* as one of the anatomical substrates in the nervous system. And since, for him, such substrates are composite containers (dense bodies with pores), the human *sensorium* is something physical that occupies a certain place in the anatomical space of the brain. Accordingly, the corporeal features of that container, as well as those of its own container, the brain, must be abstracted during the transdiction process when moving in space from local regions to all regions. And once that is done, the only features that may be generalised from the corporeal human *sensorium* to the incorporeal divine *sensorium* are two regions in infinite space, one for the cosmic spectacle and the other for the divine spectator.

As also has been learned during the course of this inquiry, Newton conceived the unknown efficient cause as a disposing principle of intellectual life (*animus*). This principle, whether human or divine, is immoveable and, hence, is extended only in the sense that it exercises its rational power of intellect to causally determine the unobservable mediators that are the unknown material causes of motion and change. Not understanding either Newton's *animus* part of his dual-soul concept or his agent-causation concept, the theologian, Jonathan Edwards, wrote in 1714, first, that 'the *Infinitum Sensorium* ... belongs to God, and *in which he moves all Bodies as he pleases, as the Soul that is in Man moves the Members of his Body*', and, second, that '*Organs of Sense and Motion* belong to God, as well as to Man' and, hence, 'God *to have a Body*, which is Unphilosophical ..., as well as Untheological'.[7]

---

[5] Buchdahl, *Metaphysics and the Philosophy of Science*, p. 596 (bold mine).

[6] Ibid., p. 596.

[7] See *supra* Pt.I.1.2, Text XII, p. 25, n.141.

That there have been difficulties in interpreting Newton's *sensorium* concept is due, in part, to Newton himself. For in his writings intended for a public audience, he neither provided a full explication of it, nor clarified its relation to his space concept. This concept, along with a number of other absolutes, made its public appearance in the first edition of Newton's *Principia mathematica*.[8] But in three of his arguments from design,[9] he complicated the problem of space by making all of space co-extensive with the divine *sensorium*. And in the General Scholium that concludes the second, 1713 edition of *Principia mathematica*, he added a further complication, when he stated that God, 'by existing always and everywhere', 'constitutes duration' or absolute time, as well as absolute space,[10] thereby bringing to the fore the problem of the ontological status of space.[11]

Newton did not treat this particular problem in the writings he made public. But in 'De gravitatione', he did explore this problem as part of his critical response to Descartes.[12] According to Ernst Cassirer, the latter had connected the problem of space with the problem of number,

> ...thereby giving the problem of space an entirely new orientation. But in the development of his metaphysics[,] space appears to be not so much a certain pattern or order beside that of number as some sort of absolute thing in the form of extended substance.[13]

That is to say, Descartes reified space as the geometrically extended impenetrable ('extension is body'). And it was precisely this reification that Newton singled out, because he believed it might lead to atheistic conceptions of matter as self-subsistent and uncreated.[14] Accordingly, he would have to find a different solution to the ontological status of space.

Since in 'De gravitatione' Newton attempts to explain why unobservable space is not a corporeal substance,[15] the question might arise: if we imagine all (real)

---

[8] See Newton, *Principia mathematica* [1687], Scholium to the Definitions, pp. 5–11, p. 5.

[9] See *supra* Pt.I.1.2, Text VIII, comment to Text VIII and Text XI.

[10] See Newton, *Principia mathematica (C-M)*, Bk. III, General Scholium, pp. 543–7, p. 545; see also Newton, *The Principia*, Bk. III, General Scholium, pp. 585–90, p. 587, where, according to the translators, the second edition had 'space, eternity, and infinity'.

[11] See Alexander, *The Clarke-Leibniz Correspondence*, pp. xxxii–xxxiii, who pointed out that solutions to the problem of space can be classed according to three questions about space: (1) the question of the ontological status mentioned above; (2) the question as to which concepts of space are most useful in physics; (3) the epistemological 'or perhaps psychological' question of how we come to acquire our knowledge of space. According to him, Clarke and Leibniz were concerned primarily with the first question; Newton, primarily with the second; whereas Kant was the first to seriously address the third question.

[12] Newton, 'De gravitatione', pp. 122–31, *et passim* (critique of Descartes), pp. 129–38 (Newton's exploration of the problem of space). Although his polemic is focused chiefly on Descartes, he also indulges in some anti-scholastic polemic, as recently has been shown by Levitin, 'Newton and Scholastic Philosophy'.

[13] Cassirer, *The Problem of Knowledge*, p. 27; for the Renaissance background to the problem of space, see also Cassirer, *The Individual and the Cosmos*, pp. 174–86,

[14] See Newton, 'De gravitatione', pp. 142–3.

[15] See ibid., pp. 133–4.

objects abolished, will the space that contains them still remain?[16] To answer this question, it would seem that the only ontological alternative to Descartes would be to reify space as an incorporeal substance. And this alternative has been suggested by Zvi Biener who, in reconsidering the arguments in 'De gravitatione for differentiating corporeal substance from space,[17] pointed out that Newton established an 'incorporeality criterion' concerning spatial 'regions of *nil* resistance [that] are, by definition, incorporeal', because those regions do not contain bodies.[18]

In Newton's day the expression 'incorporeal substance' was in common use, despite the warning of Hobbes that it was a contradiction in terms and, hence, a prime instance of insignificant speech.[19] Nevertheless, to achieve a conception of space opposed to that of Descartes, commentators have pointed to some other alternatives which (loosely) might be considered as 'incorporeal'. For example, it might be possible to reify space as a mathematical object. This was suggested by Max Jammer, who argued that a 'realistic conception of mathematics is of the first importance' not only for Newton's absolute space, but also for his methodology, 'showing, as it does, that the primary concepts underlying the structure of Newton's dynamical system are not hypothetical and unreal, justified only by subsequent experimental verification'.[20] It is worth noting, therefore, that earlier in his study Jammer pointed out that the atomists—'the very founders of the great materialistic school in antiquity'—introduced 'a new conception of reality by distinctly stating that a thing might be real without being a body'.[21]

Another alternative would be to reify space as an ideal object (hence, incorporeal) with reference to which a mathematically-inclined natural philosopher could explain his judgments and procedures. Something like this alternative was suggested by Nicolò Guicciardini, who found that he could not 'pigeonhole' Newton's mathematical method within any realist, constructivist or other meta-mathematical tradition.[22] Instead, he argued that Newton was forced to introduce absolute space in order 'to guarantee an ontological content to the forces deduced from trajectories'. And since the 'phenomena from which forces are deduced are accelerations measured as deviations from inertial motion, [it] was essential for Newton that such

[16]According to Margenau, *The Nature of Physical Reality*, pp. 253–5, this is the 'time-honored question'. Note, therefore, that this question was posed as a thought experiment by the ancient sceptic, Sextus Empiricus; see Stead, *Doctrine and Philosophy in Early Christianity*, XX, p. 310. For Newton's copy of the 1621 *Opera* of Sextus; see Harrison, *The Library of Isaac Newton*, (No. 1503), p. 237.

[17]Biener, '*De Gravitatione* Reconsidered', pp. 12–7.

[18]Ibid., pp. 13, 15.

[19]See Hobbes, *Leviathan*, p. 108; see also pp. 170–1, 428–42, *et passim*.

[20]Jammer, *Concepts of Space*, p. 95

[21]Ibid., p. 11; for a summary account of the representation of Epicurean positions in Lucretius, *De rerum natura*, see also pp. 8–11. For Newton's copy of this book, see Harrison, *The Library of Isaac Newton*, (No. 1113), p. 196.

[22]Guicciardini, *Isaac Newton*, pp. 313–5.

deviations be measured as a function of absolute time and space (in modern terms, in an inertial reference frame)'.[23]

Guicciardini's interpretation suggests that the two absolutes, time and space, function as an unconditioned standard intended to correct the appearances in the intellect by using reason to guide procedure. According to Mary and Jim Tiles, this insight is traceable to Plato, who in the *Republic* sought to distinguish philosophical expertise from opinion (*doxa*), by basing understanding (*epistêmê*) on abstract standards (*paradeigma*) by which concrete practices should be judged.[24] He spoke of these standards as if they were objects which can be identified and described in such a way that the descriptions will provide the very principles to which philosophers must appeal when justifying their procedures or judgments. Indeed, he was led to this assumption by his knowledge of the procedure by which geometricians arrive at their key terms.[25]

Finally, without multiplying alternatives, it might be possible to reify absolute space as the fundamental concept (hence, incorporeal) of Newton's dynamical system. This was the suggestion of Ori Belkind, who sought to 'deflate substantivalist readings' that treat Newton's belief in absolute space as 'a metaphysical claim about the nature of space'.[26] To achieve this aim, he provided detailed analyses of Newton's argument for absolute space, as well as his doctrine of place. He included the latter, because commentators had 'not read [that doctrine] closely enough' and, hence, had misunderstood Newton's overall argument.[27] By means of his analyses, Belkind demonstrated convincingly that 'Newton's belief in absolute space is not an idle metaphysical speculation, but an attempt to provide a solid conceptual foundation to a scientific theory'.[28]

As for the metaphysical claim deflated by Belkin, this is the claim that space consists of a continuum of substantival points, a claim first made by Samuel Clarke and from the 1980s reiterated by some commentators as Newton's position.[29] But before Belkin, different metaphysical claims have been deflated by other arguments.

---

[23] Ibid., p. 320.

[24] See Lee (tr.), *Plato The Republic*, [Pts. VII and VIII], Bks.VI and VII, especially the epistemology represented in the similes of the sun, the divided line and the cave, which, in short is that the intellectual eye, fixed on objects in the twilight world of becoming, only form opinions and seems to lack intelligence; whereas the same eye, fixed on objects illuminated by reality (i.e., *paradeigma*) in the world of being, understands and possesses intelligence (i.e., knowledge).

[25] See Tiles and Tiles, *An Introduction to Historical Epistemology*, pp. 16–23, *et passim*. Newton himself owned a copy of the 1626 Latin translation of Plato's *Republic*, and he also owned a copy of the 1602 copy of Plato's works in Greek and Latin as interpreted by Ficino; see Harrison, *The Library of Isaac Newton*, (Nos. 1324–5), p. 218.

[26] See Belkind, 'Newton's Conceptual Argument', p. 274.

[27] Belkind, 'Newton's Conceptual Argument', p. 271; see also pp. 276–80 for his interpretation of the doctrine of place. Note that Newton defined place as *part of space*, i.e., as the volume a body occupies within a larger volume that contains it. He thus rejects Descartes's definition of place (as Belkind pointed out), as well as Aristotle's definition summarised *supra* Pt.IV.4.3, p. 119.

[28] Ibid., pp. 290–1.

[29] See Ray, *Time, Space and Philosophy*, pp. 108–10, 146–50, 215–6.

Consequently, there has been a continuing pattern of denial and assertion concerning Newton's space concept. For example, Stephen Toulmin considered that Newton's scientific theory concerning space stood on its own and was vitiated by 'metaphysical accretions' that should not be taken literally,[30] whereas in response to Toulmin, Wolfgang von Leyden pointed out that

> ...what Newton did was not to stretch a scientifically defined term like that of space beyond the range of its physico-mathematical application, but to add *to its more limited scientific use a set of wider metaphysical assumptions* ... [in order] to give his theory a metaphysical foundation or at least a link with metaphysics.[31]

From what has been said in the Introduction and in Part IV of this inquiry, Newton neither added metaphysical accretions (Toulmin) nor a metaphysical foundation (Leyden). Rather, his remarks concerning God as both first and final cause are tenets of faith and as such, are separable from natural philosophy.

Indeed, there is absolutely no hint of an identity between God and space in the writings Newton made public, even though he conceived both entities as unobservable incorporeal substances that are immovable (eternally fixed). But space is inert and, hence, causally inefficacious, whereas God, as the impersonation of intelligence, is not only the unknown first and final cause but also Newton's functional principle. Situated in the enormous vacuum beyond the system of the fixed stars, the divine spectator exercises his all-seeing spectatorial eye of intellect to look down upon the spectacle in the region of cosmic space in order to sustain the material cosmos by causally necessitating 'at will' the motion and change of cosmic bodies. But God himself does not move, because his volitions are carried out either directly, by unknown efficient causes,[32] or indirectly, by unknown material causes (mediators) and by the ordained laws (to be discovered) that God implanted in the bodies of the solar system and the nervous system.

Although Newton specifically states that God's incorporeal substance is unknowable, yet in a passage from Text XII, Newton reminds his readers that God does not have parts, because he is a 'uniform being', that is to say, the uni-form that gives all forms their being.[33] And in an undated manuscript, he asserted this explicitly at the beginning of his long definition of the 'most perfect idea of God', namely, 'that he be *one substance, simple*', '*a substance* which by his own presence discerns and rules all things, just as the cognitive part of man [intellect] perceives the forms of things brought into the brain, and thereby governs his own body'.[34]

---

[30] See Toulmin, 'Criticism in the History of Science'.

[31] Leyden, *Seventeenth Century Metaphysics*, pp. 229–60, pp. 256–7. The context of his argument suggests that, for him, a 'metaphysical foundation' is synonymous with a 'theological foundation'.

[32] I.e., not by miracles; see *supra* Pt.IV.4.2, p. 106 and n.25.

[33] See *supra* Pt.I.1.2, Text II, p. 26 and nn.147, 148.

[34] See Newton, 'Tempus et Locus', p. 123, §7 (italics mine). The translator from the Latin suggested that the manuscript dates from the early 1690s, but he provided no supporting evidence for this.

According to Christopher Stead,[35] this doctrine was derived not from Scripture but from mathematically-minded philosophers such as Pythagoras and his followers, who supposed that just as number 1 was the source of all cardinal numbers, so too the 'monad' was the source of all rational order in the universe or, conversely, that the creative power behind that order had the characteristics of the 'monad'.[36] In Newton's day, it was not unusual to conceive the principle of intellectual life in humans or its rational knowing power of intellect, as a simple (indivisible) incorporeal substance,[37] for according to Charles Morris, the substance conception of mind is the oldest theory,[38] evidence for which is found in some fragments of the pre-Socratic philosophers.[39]

Although Newton's substance conception of *animus* appears to have been common in his day, yet it is probable that his functional conception of *animus* as both spectator and causal agent is indebted to Plato, as well as Descartes, for they provided the two classic but slightly different expressions of the rational soul as an incorporeal substance. According to Plato, this substance differs in essence from physical nature, not as a spectator, but as the principle of order and movement in the cosmos. But he never made intelligible the relation of the principle of order and movement to the ordered and moving world or the relation of thought to the other constituents of reality.[40] According to Descartes, the soul is indeed different in essence from physical nature, though not as a principle of order and movement, but rather as a spectator that reads the images transmitted to, and represented inside the brain.[41] Moreover, to this dualism he added another between the rational and the sensuous functions,

---

[35] See Stead, *Doctrine and Philosophy in Early Christianity*, I, pp. 180–1. Other articles in his collection consider whether the doctrine implies that God cannot have properties but can still exercise a plurality of functions and whether the doctrine is a problem for Christian orthodoxy.

[36] OED 1: *Now* applied only with reference to the Pythagorean or other Greek philosophies, in which numbers were regarded as real entities and as the primordial principles of existence; see also 1b (applied to the deity), where the first two citations are to the Platonists, Henry More and Ralph Cudworth (the different usage of Leibniz is given afterwards).

[37] For Willis's conception, see *supra* Pt.III.3.2, p. 77. Note that in 1705 and 1712 Samuel Clarke, promulgated a different version of the doctrine of God as absolutely simple in opposition to the doctrine set down in Article 1 of the 'Thirty-nine Articles', subscription to which was part of the requirement for membership in the Church of England; see Kassler, *Seeking Truth*, pp. 205–6 and, for a response to Clarke, pp. 291–2.

[38] See Morris, *Six Theories of Mind*, who indicated, p. 16, that Aristotle was the earliest opponent of the substance conception of mind, even though 'the substance approach continually creeps into' his account.

[39] See Kirk, Raven and Schofield, *The Presocratic Philosophers*, p. 366, and, in particular, the variety of dualism evident in the fragments of Anaxagoras of Clazomenæ, pp. 363–6, who described mind, 'like everything else,' as 'corporeal', but claimed that mind 'owes its power partly to its fineness, partly to the fact that it alone, though present in the mixture [of matter], yet remains unmixed', i.e., pure.

[40] See Morris, *Six Theories of Mind*, pp. 12–20.

[41] See Kassler, *Inner Music*, pp. 44–8.

because his mechanistic physiology of reflex did not permit him to treat the rational soul as a causal factor that would account for the interaction of mind and body.[42]

It is not surprising, therefore, that in Newton's attempt to retain *animus* as a causal factor, he seems to have combined the two classic conceptions of Plato and Descartes.[43] But in so doing, he located *animus* in an undisclosed place in the anatomical space of the brain, thus not only confining it to its own subjective perceptions but also restricting it to the intellectual activity of knowing. This kind of intellectualism, which Gilbert Ryle had called 'the para-mechanical theory of mind',[44] leads to difficulties not only in conceiving the causal relation of God to the processes of nature, but also in conceiving the relation of the human intellect to the human body.[45] In other words, there is a gap between the spectacle and the spectator.[46]

Indeed, the source of this gap is Newton's substance realism, with its thesis that entities to which ontological commitment is made exist independently of our knowledge of them. This kind of realism occurs when abstract terms are used substantively, that is to say, when they are mentally converted into things or real entities, each with its absolute, though empirically indiscernible meaning.[47] Newton's commitment to space as real (absolute) is an example of this mental conversion, and so is his commitment to other unobservables, including incorporeal *animus* and *anima*, which he assumes is more real than the experiential content of, or appearances represented in the *sensorium*. As Morris pointed out long ago, the paradox of English empiricists is that their substance realism became 'ensnared in the toils of a

---

[42] See ibid., pp. 21–30.

[43] According to Yolton, *Thinking Matter*, p. 107, for those who found real causation only in God, 'the result was that God was active in the world, either at the beginning or continuously'.

[44] See *supra* Pt.IV.4.2, p. 108.

[45] In the words of Morris, *Six Theories of Mind*, p. 42: 'how could a mind existentially confined to its own states know a world different in kind from itself?' For Edwin Burtt, the relation of 'man' to nature was a central problem for the science that developed from the time of Copernicus to the time of Newton; in addition to many references to this problem in the main part of his text, see Burtt, *The Metaphysical Foundations of Modern science*, pp. 310 *seq.*

[46] In classical (Newtonian) physical theory, this gap occurs, because the measurement process is fully analysable in terms of the equations of motion alone, so that the interaction between object and observer is negligible; for an instance of the gap in applied physics, see Kassler, 'Musicology and the Problem of Sonic Abuse', pp. 328–9. But in the new physics (quantum and relativity), there is an unanalysable element in every measuring process that is due to the intervention of the observer on the object. Consequently, since the observer can no longer be regarded as an innocent registrar of his or her objective observations, predictions can never be exact and only probability values can be established.

[47] In Newton's day, there were two rival definitions of the term 'substance': one as that in which properties ultimately inhere but which itself is not a property or quality; the other as that which is independent or self-sufficient, not requiring anything else in order to exist. Roger North followed the first definition, whereas Samuel Clarke followed the second. Therefore, they may be described, respectively, as epistemological and ontological realists; see Kassler, *Seeking Truth*, pp. 184–5 and nn.299–302.

subjectivistic metaphysics'.[48] And in the case of Newton, this leads to a further paradox, for in spite of his attempts to overthrow the mechanical philosophy of Descartes, he retained the dualism that is a fundamental feature of that philosophy.[49]

Of course, all three of Newton's concepts are theoretical constructs;[50] and, as Henry Margenau warned, 'Invention of a construct does not carry with it the assurance that the construct is scientifically acceptable or that it is a part of reality'.[51] So, we might ask, what happened to the three concepts during the course of history? Of the three, Newton's space concept has dominated the attention of commentators. Accordingly, the reasons are now well known why sometime after the end of the nineteenth century, this concept ceased to be considered as part of physical reality. The timetable of its demise may be summarised as follows: by the middle of the nineteenth century, 'it became clear that the concept of absolute space was useless in physical practice'; towards the end of that century, 'it became obvious that absolute space evaded all means of experimental detection'; and by the turn of the century, not only mechanics but also physics were 'ready to abandon the concept of absolute space altogether'.[52]

As for the *animus* part of Newton's dual soul concept, evidence has been drawn chiefly from two contexts: his arguments from design and his consideration of the problem of muscle action. In the former context, his statements about its substance are overt; but in the latter context he used the term 'soul' only once to mean an in entity in the body that is an unknown efficient cause, the meaning of which is elucidated by the examination in Part IV of Newton's agent-causation concept. According to that concept, an agent (efficient cause), whether divine or human, exerts its power of intellect, which then mediates the will, understood as an agent's intellectual preferences (volitions) in regard to action. It should be noted, therefore, that according to Charles Morris, there have been two tendencies in Western thought, one of which identifies intellect ('mind') as soul and the other of which makes intellect the thinking 'aspect' of soul.[53] It is the latter tendency that seems to be represented in the writings of Willis, Boyle, Locke and Newton, since each distinguished between a disposing principle (*animus*) and its disposing power (intellect). Although this is a subject for further investigation, it is important to recognise the terminological problems raised by too easy recourse to the term 'mind'.

During the period of this inquiry, the incorporeal substance concept remained dominant, even though a materialistic variant of the substance concept gradually came to the fore.[54] Nevertheless, according to data presented by Edmund Boring

---

[48] Morris, *Six Theories of Mind*, p. 38.

[49] See ibid., p. 27.

[50] Perhaps they belong to a class of hypotheses that Gerald Holton called 'thematic' and described as 'precisely built as a bridge over the gap of ignorance'; see Holton, 'Presuppositions in the Construction of Theories', pp. 91–3.

[51] Margenau, *The Nature of Physical Reality*, p. 73.

[52] See Jammer, *Concepts of Space*, pp. 138, 138, 148.

[53] See Morris, *Six Theories of Mind*, p. 10.

[54] See Yolton, *Thinking Matter*, who traces aspects the immaterialist-materialist controversy in the eighteenth century.

and by Edwin Clarke and L. S. Jacyna, the incorporeal nature of governing processes continued throughout the nineteenth century.[55] Even so, these three commentators regarded various versions of the incorporeal substance concept as deleterious to scientific advance and, the latter two commentators, in particular, regarded that concept as a hindrance to any idea of 'a site or location for mental phenomena'.[56] In doing so, however, they ignored the difficulties in elucidating the concept of mind itself. Morris treated some of these difficulties when analysing five theories that by 1932 had succeeded the doctrine of mind as substance.[57] According to him, since all five theories were 'at one in their opposition to an explanation of mind as substance',[58] he concluded that a 'fair interpretation of the verdict rendered in the court of world history' is that 'the category of substance has outlived its metaphysical usefulness'.[59]

Last but not least, there is Newton's *sensorium* concept, part of which falls under the problem of the corporeal human *sensorium*, and part, under the problem of the incorporeal divine *sensorium*. There has been no historical treatment of the 'after life' of each part of his concept. And the reason for this is that the foregoing inquiry represents the first systematic investigation of his concept as a whole. However, it seems probable that neither part of his concept made a major impact on later generations, although this probability remains to be thoroughly tested.[60] Nevertheless, as the data provided by Boring, as well as Clarke and Jacyna make clear, the *notion* of a human *sensorium* continued long after Newton's death.[61]

For example, in the latter part of the eighteenth century, Albrecht von Haller conceived the entire white matter of the brain as a *sensorium commune* and as the seat of *all* sensations (not just those derived from the external senses).[62] By remaining indebted to Galen's conception of the brain as a unity and as the sole center of

---

[55] See Boring, *A History of Experimental Psychology*, who focused on the period 1860–1910, though with attention to preceding developments; see also Clarke and Jacyna, *Nineteenth-Century Origins of Neuroscientific Concepts*.

[56] Ibid., p. 275.

[57] The five theories conceived mind, respectively, as process, act, relation, substantive and function.

[58] See Morris, *Six Theories of Mind*, p. 28.

[59] Ibid., p. 46.

[60] Insofar as the divine *sensorium* is concerned, that part of Newton's concept was not mentioned by William Paley in his book-length argument from design. In addition, in a single passage on the human *sensorium*, he made no mention of Newton, when he wrote: 'if there be those who think, that the contractedness and debility of the human faculties which the expectations of religion point out to us, I would ask them, ... what advancement and improvement, the rational faculty ... may not admit of, when placed amidst new objects, and endowed with a *sensorium*, adapted as it undoubtedly will be, and as our present senses are, to the perception of those substances, and of those properties of things with which our concern may lie'; see Paley, *Natural Theology*, pp. 595–6 (italics mine).

[61] See Boring, *A History of Experimental Psychology*, pp. 36, 75–6, 82–5, 87, 101, 143–4, 668, 672–4, 714; see also Clarke and Jacyna, *Nineteenth-Century Origins of Neuroscientific Concepts*, pp. 92, 105–6, 125–6, 215–6, 330, 334, 403 n.97, 427 n.17. Neither source mentions Newton's *sensorium*.

[62] Ibid., p. 105.

the nervous system, all other structures being derivative, he ignored Willis's modification of this ancient conception by introducing the cerebrum and cerebellum as a dual control system.[63] By contrast, Jiří Procháska believed that a sensory impression is converted by a reflex action into a motor message in a *sensorium commune* of the spinal cord, brain stem, cerebral and cerebellar peduncles and parts of the thalamus, thereby adopting the emerging conception of the brain as a unity in multiplicity.[64]

In the nineteenth century the location of a *sensorium commune* in a reflex centre of the spinal cord became controversial. For example, Marshall Hall believed that there was a property of the sensory and motor system of nerves which was independent of sensation and volition, whereas Johannes Müller retained a *sensorium commune* in the spinal cord, because he could not agree that all reflex action occurred without sensation.[65] In the mid-nineteenth century Robert Remak claimed that just as a conscious *sensorium commune* centralised in the cerebrospinal axis regulated animal functions, so too an unconscious *sensorium commune* centralised in the vegetative nervous system regulated vegetative functions.[66] And from the 1860s attempts to localise function in the white matter of the brain would gradually disappear when cerebral localisation in the grey matter or cortex 'became a scientific reality'.[67]

This brief review of the afterlife of Newton's three concepts may suggest the need for further research, especially in regard to the long history of the idea of a human *sensorium*. Unfortunately, however, future investigations into the post-history of a *sensorium commune* will face considerable difficulties in tracing changes to Willis's common sensory concept as constituted by two anatomical substrates in the middle of the brain (cerebrum), namely, the corpora striata and corpus callosum. These difficulties are due in part to two tendencies among commentators who, on the one hand, treat the progressive (whiggish) aspects of its history, and, on the other hand, utilise modern anatomical terms to elucidate this history.[68] It is worth pointing out, therefore, that at a date still to be discovered, Willis's term 'corpora striata' was superseded by the term 'corpus striatum', which refers to part only of

---

[63] See *supra* Pt.III.3.2, p. 82.

[64] Clarke and Jacyna, *Nineteenth-Century Origins of Neuroscientific Concepts*, pp. 103–6. Note that Procháska claimed to have modelled his procedures on certain aspects of Newton's scientific procedure; that he was not alone in this claim; see *supra* Pt.II.2.2, p. 50 n.108.

[65] Ibid., pp. 116, 125. For an extensive treatment of Müller's theories, see Boring, *A History of Experimental Psychology*, pp. 82–90, *et passim*.

[66] See Clarke and Jacyna, *Nineteenth-Century Origins of Neuroscientific Concepts*, p. 92. Nevertheless, in 1959 one well-known neuroscientist returned to the notion, but not the term, of a *sensorium commune* in the oblong marrow (medulla oblongata); see Kassler, *Music, Science, Philosophy*, pp. 290–1 and n.60.

[67] See Clarke and Jacyna, *Nineteenth-Century Origins of Neuroscientific Concepts*, p. 303 (italics mine).

[68] In the book cited in n.67 above, there is some evidence of these two tendencies, especially in regard to the 'corpus striatum', for this is the commentators' usual term (even for Willis) and so indexed, even though, pp. 248, 263, two of the writers surveyed by them used the term 'corpora striata'. For an instance of this problem, see also *supra* Pt.III.3.2, p. 84 n.123.

the basal ganglia, namely, the caudate and lenticular nuclei together with the internal capsule that separates them. As for his term 'corpus callosum', this is still in use, but it now refers to the gross transverse commissure, or broad band of white matter at the bottom of the longitudinal fissure of the cerebrum, that connects the cerebral hemispheres.

Regarding the questions posed in Part II concerning Newton's *sensorium* concept, for which only plausible or probable answers could be given in Part IV,[69] it may or may not be the case that as Newton's unpublished writings come under closer scrutiny, so too more information will emerge concerning changes to, or developments of the whole or parts of his concept. And it also may or may not be the case that some information will come to light that could provide an answer to a question that falls outside the remit of the foregoing inquiry. The question is this: What were the motives that prompted Newton to investigate, as well as formulate his *sensorium* concept as a theoretical construct?

---

[69] See *supra* Pt.II.2.2, p. 57 and Pt.IV.4.1, pp. 99–104.

Appendix

Isaac Barrow's Lecture X from the Second 1665 Set of his Lucasian Lectures on Mathematics[1]

In Lecture X Barrow raised two related problems: space (in its real or absolute sense) and infinity.[2] These problems had long been regarded as *insolubilia* and, hence, as ontological paradoxes. But the two problems in question, like most paradoxes, also challenge some orthodoxy. For example, space, which is a *nihil* paradox, is 'no Thing' because unobservable; and this paradox was closely allied with the Christian doctrine that God created the world, as well as space and time, *ex nihilo*. But Barrow not only asserted that space is *not* 'a mere Nothing', but also that its infinite extent is eternal and uncreated, because it existed 'before any Beginning'.[3] And he was not alone in holding this heterodox opinion, for, amongst others, it was also held by Pierre Gassendi,[4] J. B. van Helmont, Henry More and, later on, by Newton himself.[5] And even before such an opinion became heterodox, it was also held by Plato, who wrote:

---

[1] Barrow's second set of Lucasian lectures were delivered in 1665, and this set along with the other sets were published in 1683 as *Lectiones mathematicæ*; see Feingold, 'Isaac Barrow: Divine, Scholar, Mathematician', pp. 68, 79. The extracts of the lecture reproduced here are from the English translation in Barrow, *The Usefulness of Mathematical Learning*, pp. 163–85.

[2] In the previous Lecture IX Barrow explored the logical foundations of infinitesimals. According to Malet, 'Barrow, Wallis, and the Remaking of Seventeenth Century Indivisibles', pp. 75–81, Barrow upheld the thesis that magnitudes are actually composed, or consist of, an infinite number of infinitely small parts; and by assuming that space itself is infinite, one can produce innumerable instances of 'an infinite number within an infinite number'. Lecture IX, therefore, leads to the two ontological problems treated in Lecture X.

[3] See ibid., pp. 181 and 170.

[4] Although Gassendi postulated an infinite God, yet because he wished to avoid the radical natural-philosophical conclusions of Cusanus and Bruno, he tended to be wary of infinitude in any form; see Brush in Gassendi, *Selected Works*, p. 382.

[5] The name of Samuel Parker was added to this list by Roger North; see Kassler, *Seeking Truth*, p. 186 n.306.

J. C. Kassler, *Newton's* Sensorium*: Anatomy of a Concept*, Archimedes 53,
https://doi.org/10.1007/978-3-319-72053-1_6

> ...there exist [form, copy and] ...space which is eternal and indestructible, which provides a position for everything that comes to be, and which is apprehended without the senses by a sort of spurious reasoning and so is hard to believe in—we look at it indeed in a kind of dream and say that everything that exists must be somewhere and occupy some space and that what is nowhere in heaven or earth is nothing at all.[6]

The problem of infinity was regarded as beyond the limits of thought and, hence, as incomprehensible. But its complexity was due in part to the state of mathematics regarding infinity, as well as from the fact that speculation into 'nothing' was usually undertaken in conjunction with notions of space and God as infinite. According to Rosalie Colie,

> ...the appropriate way to express transcendent deity is by tautology (e.g., God's own self-referential comment to Moses "I am that I am"[7]) or by negative affirmations which are by definition paradoxical (e.g., God is incomprehensible, or infinite). That is, there are two ways of avoiding contamination of the divine essence by the implications carried in metaphors from human experience, the one, to use terms of totality about divinity (Omniscience, Omnipotence, Omnipresence); the other, to use negative or "ablative" terms (Infinity, Eternity, Immutability).[8]

In his lecture Barrow resorted to both kinds of terms for God, at the same time attempting to bring the problems of space and infinity within the domain of mathematical reasoning and knowledge.[9] Consequently, the aim of his lecture is to provide a method of conceiving space that is concordant with faith, ontology and mathematics, particularly geometry. It is worth noting, therefore, that in a section of his lecture, Barrow attempted to provide a brief geometrical demonstration that space is real and also distinct from magnitude or quantity of matter, so that, contra Descartes, the latter is *not* dispersed everywhere.[10] It is also worth noting the probability that Barrow's lecture led Newton to recognise that the mathematisation of space was a prerequisite for what we now call 'mathematical physics'.[11]

In the extracts from the lecture that follow, the original page numbers are inserted between square brackets in the text itself. The same procedure is used for Barrow's footnotes, which in the original source are identified by alphabetical letters. This

---

[6] See Lee (tr.), *Plato Timaeus*, pp. 71–2.

[7] The Second Book of Moses called Exodus iii.14: 'And God said unto Moses, I AM THAT I AM'.

[8] Colie, *Paradoxia Epidemica*, p. 25, also for some early modern solutions to the ontological paradoxes of being, becoming and creation, pp. 145–68, 219–51 and 300–28. For a more complete treatment of pp. 300–28, see Colie, 'Some Paradoxes in the Language of Things'.

[9] In this mathematical emphasis, Barrow differed (as does Newton) from Gassendi, Helmont and More, although they all held that space is eternal and uncreated, as noted above.

[10] See Barrow, *The Usefulness of Mathematical Learning*, pp. 171–2. Note that Samuel Clarke may have reproduced this demonstration in a letter to Roger North which is no longer extant but which was paraphrased by North in his reply to Clarke; see Kassler, *Seeking Truth*, pp. 130–39, pp. 133–4.

[11] For his 'significant characterisation of three-dimensionality of space' in *Principia mathematica*, see Bochner, *The Role of Mathematics in the Rise of Science*, p. 249. For Barrow's impact on Newton during the latter's formative years, see Feingold, 'Newton, Leibniz and Barrow too', pp. 312–24.

modification to the extract then allows footnotes to be identified by numbers and to be used for brief comments relating to the lecture. When Barrow's meaning requires clarification, this is provided either in a footnote or in the text itself by the insertion of a word in square brackets. Citations from sacred writings are given from the King James Version of the Bible; but Barrow's references to Aristotle's *Physica* are left without comment.[12]

Lecture X

Of Space and Impenetrability

SINCE the Thread of my Discourse has led me, for the sake of Mathematical Hypotheses, to treat of the more common Affections of Magnitude; and since I have already spoke something (little indeed considering the Dignity of the Subject, but enough for our Design) concerning some Things which first offer themselves, *viz.* the *Termination, Extension, Composition, and Divisibility* of Magnitude; it now follows that I should touch upon what remains, *viz.* its *Occupation of Space, Determinate Position, Mobility, Mensurability, Proportion,* and whatever else occurs from whence Mathematicians do, or lawfully may, draw Hypotheses suited to their Ratiocinations.

In the first Place, it is wont to be attributed to Magnitude, that it *occupies,* or *fills Space.* But it is difficult to explain what this *Space* is. For if we either attend to vulgar Conceptions, or examine the [**164**] more subtle Sentiments of Philosophers, it is not easy to determine whether there really is any *Space* distinct from Magnitude or not; the Opinions on each side seeming both to depend upon very specious Arguments, and be pressed with weighty Difficulties.

On the one Hand, these Arguments seem to make it sufficiently plain, that there is really no such Thing as *Space* existing separate from the Quantities of Things. And that first, because, if it be unproduced and independent, it must consequently be also eternal and immense (for were it not (say they) that such a Thing is for the most Part conceived and supposed by the Assertors of real *Space*; if such a Thing be not, then every Reason will fail upon which it is supposed, as will appear to any one who weighs the Causes of framing it;) but for any Thing to partake of the singular Attributes of the Divine Nature, such as not to be created nor dependent upon God, seems contrary, as well to right Reason, as Religion. Again if we examine the Idea of *Space* whatsoever it be, we seem to apprehend nothing by it but a certain Extension and indefinite Capacity; which, since they are Properties of Magnitude itself, do argue no Difference of Space from Magnitude; for why should that differ in the Thing, which agrees in the Properties? And this is not only the Argument of *Cartesius*, but of *Aristotle* himself. *If* (says the Philosopher) *the Magnitude of Things*

---

[12] Barrow owned several works of Aristotle, for which see Feingold, 'Isaac Barrow's Library', (Nos. 68–73), p. 342.

have nothing in itself different from Space, then wherefore is the Bulk of a Thing distinguished from its Space? Moreover, if *Space* be any Thing different from Magnitude, we may enquire to what Classis of Things it is to be referred: For every Thing either subsists of itself, or is an Accident to another Thing; but neither of these seems to agree with *Space*. The Patrons of [real] *Space* do not exalt it to the [**165**] Dignity of a Substance, nor will the Thing itself bear it. Neither is it an Accident, because it is extrinsecal to all Substance; since it is not carried about with it, but remains, though that be taken away, and depends upon no other Thing.[13] I pass by the Device of *Zeno*:[14] That every Thing is somewhere, therefore if *Space* be any Thing distinct from other Things it must exist somewhere; whence there will be a *Space* of a *Space*, and another *Space* of this second *Space*, and so on infinitely (as *Aristotle* observes [Phys. IV. 3]) which is ridiculous. By such Arguments as these is the real Diversity of Space from magnitude opposed: but it is defended from Arguments which are equally strong in Appearance.

For *first*, if we appeal to the common Conceptions of Men, the Notion of a Space distinct from Things does seem to be either innate or acquired. *All Men* (says *Aristotle* [Phys. IV. 1]) *do in Thought separate the* Ubi *of Things from their* Esse. The Vulgar are accustomed to imagine that there is some common *Substratum* to all Things, which is infinitely extended, and circumscribed within no Limits; which is perfectly penetrable, and easily admits every Thing within itself, not resisting the Entrance of any Thing; which receives the Successions of moveable Bodies, determines the Velocities of Motions, and measures the distances of Things; which is immoveably fixed, has none of its Parts tied to any Thing, nor can possibly be any where transferred from where it is; which lastly is a Receptacle of immense Capacity, or as the Philosopher speaks [Phys. IV. 6] *an immoveable Vessel*, containing within it all the Things that either are, or can exist. Such is the Notion of *Space*, which is engraven in the Imaginations of all Mortals. And very many [**166**] Things beside seem to argue the real Existence of such a Thing, as well as this Consent of the Imagination: To make the Force of which apparent and effective, certain Positions or Assertions are to be laid down, whereby the Arguments brought for the Reality of *Space* are supported.

First then[,] Matter is not infinitely extended; at least, which is enough for our Purpose, it is not so of Necessity. For from whence can it have this Necessity? Is it from itself? This is impious, since the sacred Writings do often expresly ascribe the

[13] Note that at some point in the next decade, Newton echoed Barrow in writing that space 'has its own manner of existence which fits neither substances or accidents [of matter]', so it cannot be said 'to be nothing, since it is rather something ... and approaches more nearly to the nature of substance. There is no idea of nothing, nor has nothing any properties; but we have an exceptionally clear idea of [geometrical] extension, abstracting the dispositions and properties of a body so that there remains only the uniform and unlimited stretching out of space in length, breadth and depth'; see Newton, 'De gravitatione', p. 132 (additions in brackets mine).

[14] I.e., Zeno of Elea, whose 'Device' referred to is probably not one of his paradoxes but one or another of the extant antimonies that concern philosophical problems of infinity; see Kirk, Raven and Schofield, *The Presocratic Philosophers*, pp. 265–9, and for Plato's indebtedness to these, p. 279.

Origin of all Things to God: *All Things* (says St. *Paul* [Eph. iii. 9][15]) *were created by him and for him.* And thus sung the four and twenty Elders [Rev. iv. 11[16]], *For thou hast created all Things, and for thy Pleasure they are, and were created.* Again, as the Evangelist St. *John* expresses it [Joh. i. 3[17]], *All Things were made by him, and without him was not any Thing made that was made. For all those Things*, says the Prophet *Isaiah* [Isa. lxvi. 1[18]], *hath mine Hand made, i.e.* all Things whatsoever without any Exception: And innumerable such other Expressions occur every where in Scripture. Nor is there any Reason why we should believe Matter to have an infinite Substance necessarily of itself. Is it then from God? Who imposed this Necessity upon him who is a perfectly free and independent Agent, that he appear to have been attributed to it? I do not dispute whether or no he is able to do so, for who can assign Limits to the Divine Power? But it seems far more credible that, as he has limited the Powers of other Things, so he has appointed certain Bounds to Matter. The *holy Scriptures* do seem plainly to attest as much. *Behold* (says **[167]** the most wise King [2 Chron. ii. & vi. 18[19]], more than once) *the Heaven and the Heaven of Heavens cannot contain thee, i.e.* the whole Universe is too narrow to exist with God; he transcends the utmost Bounds of Things, and consequently Matter is not in fact extended *ad infinitum*, much less is it so of Necessity.

To this it is replied by *Cartesius* [Prin. II. 21[20]]; *It is necessary for Matter to be infinitely extended, for this Reason, because we conceive its Bounds everywhere; we do not only imagine certain Spaces to be extended indefinitely beyond ourselves, but they are moreover truly imaginable*; *i.e. we perceive that they are real, and consequently that a bodily Substance indefinitely extended is contained in them*; *since the Idea of that Extension, which we conceive to be in all Space, is altogether the same with the Idea of a bodily Substance.* But this Argument seems to contain in it more

---

[15] The Epistle of St. Paul, the Apostle, to the Ephesians iii.9: '...who created all things by Jesus Christ.'

[16] The Revelation of St. John the Divine: iv.11: '...for thou hast created all things, and for thy leisure they are and were created.'

[17] The Gospel according to St. John: i.3: 'All things were made by him; and without him was not any thing made that was made.'

[18] The Book of the Prophet Isaiah: lxvi.2 (not, as above, 1): 'For all those *things* hath mine hand made....'

[19] The Second Book of the Chronicles: ii.6: '... seeing the heaven and heaven of heavens cannot contain him? '; vi.18: '...behold, heaven and the heaven of heavens cannot contain thee....' For other similar passages, see The Fifth Book of Moses called Deuteronomy x.14, and The First Book of the Kings viii.27.

[20] I.e., *Principia philosophiæ*. See Descartes, *Principles of Philosophy*, Pt. II, §21, p. 49: '...we understand that this world, or the universe of material substance, has no limits to its extension. For wherever we may imagine those limits to be, we are always able, not merely to imagine other indefinitely extended spaces beyond them; but also to clearly perceive that these are as we conceive them to be, and, consequently, that they contain an indefinitely extended material substance. Because (as has now been shown at length), the idea of that extension which we conceive in any space whatever, is exactly the same as the idea of material substance.' First published in Latin in 1644, then in French in 1647, Barrow owned a copy of the first Latin edition; see Feingold, 'Isaac Barrow's Library', (No. 321, p. 349.

of Subtilty than Solidity. For *first*, no Body either does or can conceive Matter to be actually infinite; but to conceive it to be indefinitely extended, is the same as to imagine it to have no Terms or certain Bounds; as the vulgar Part of Mankind do think the Surface of the Earth to be infinitely extended; or as one standing upon the Shore thinks the Sea to be encompassed with an uncertain Bound; or as we conceive the Grains of Sand of the Sea to be indefinitely many. Moreover because we can imagine certain Spaces beyond any determinate Bounds; it by no means follows that Matter does actually exist beyond. However it may be probably gathered that it therefore may exist, because whatsoever we conceive as evidently possible, that may be effected by the Divine Power. But who in his right Mind will deny, that we may imagine innumerable Things, which neither are nor ever will be? Nothing is repugnant, and [**168**] we may easily imagine with the vulgar, that the Earth reaches to the utmost Bounds of Heaven and the farthest Limits of the Universe: we may suppose in our Minds the Sun to be a thousand Times greater, the Moon many Parasangs[21] nigher; the Stars vastly more numerous, and innumerable other Things, which are by no Means impossible, nor altogether absurd; but are these Things therefore true in fact? No certainly, they are no more than so many Dreams or Deliriums of sick Men. Imaginability then does at most only prove the Possibility of a real Thing, but no where its actual Existence. The World is not proved to be actually infinite from the Imagination, but only to be greater in Power than any finite, (*i.e.* determinately finite) Thing.

But to be satisfied how very much *Cartesius's* great Subtilty has failed him in this Case, let us search and weigh a little the Origin of this our Imagination of *Spaces* beyond the World, and we will certainly find it like most other Things, to be derived no otherwise than from the Senses.[22] For since we scarce attain any Thing by any Sense, but we still experience something equally sensible by proceeding beyond it, especially when lifting up our Eyes to Heavens we behold the vast Chasm on every Side contained within no perceptible Limit, but running forth into Regions unknown to us; hence an Occasion is offered of describing in our Fancy a certain immense or indeterminate Gulph of heavenly *Space*, in which the Clouds are suspended, the Winds blow here and there, and the Stars seem to swim like so many Fishes: from which Sensation notwithstanding, or Imagination, it seems very absurd to infer any Thing concerning the true Extent of the Universe, whether it is finite or infinite. For every Sensation is of Things particular, and from the Existence of a particular Thing, we may conclude that something like it may exist, but not that it actually does exist, as has been sometimes taken notice of before.

I say nothing as to the Parity of Reason in attributing *Eternity* and *Independence* of Matter, as well as a *necessary Infinity of Extension*; for as there is some Space

[21] OED: A Persian measure, usually reckoned as equal to between 3 and 3½ English miles (1555–1882).

[22] In a manner appropriate to an audience of mathematical students, the remainder of the paragraph constitutes a brief instance of generalising from the observable (known) to the unobservable (unknown), a move that is central to Newton's more developed method of reasoning from experience.

beyond any Limits of the Universe, so has it been before any Beginning. We are wont to have the same clear Imagination of some Time after the End of any Time, and consequently by this way of arguing, Matter may be demonstrated to be necessarily eternal, whence it follows that it is also independent: which Properties notwithstanding of the Divine Perfection, it is most dangerous to attribute to any other than the great God himself; and is in the highest Degree to be shunned by a Christian Philosopher. It may be therefore taken for granted, that Matter or Body is not altogether unbounded; at least it is not so necessarily. Let this be first supposed.

Let it also be assumed, *secondly*, that it is in the Power of God to augment or diminish the Matter that now exists, according to his Pleasure, *i.e.* to create, or reduce to Nothing, what Portion of it he pleases. Faith commands, and Piety compels us to admit this; nor does Reason oppose, but rather defend and confirm it. For since we can conceive Matter to be larger than any Thing determined,[23] therefore nothing hinders, nay it evidently follows from thence, that God can cause it to be so. But if it may be enlarged by the same Power it may be diminished; and the Web which he hath newly wove, he can with the same Ease unravel. And because we can imagine Matter more contract, nay suppose it Nothing at all, it is therefore in the Divine Power to perform it.

[**170**] Moreover, *thirdly*, it is most easy to conceive, and ought by no Means to be denied, that God can preserve every Thing in its present State and Situation, so as not to be intrinsecally changed by any extrinsecal Accidents, much less to have its Nature entirely destroyed; *viz.* that a Right Line, a plain Superfice, the Circumference of a Circle, the Roundness of a Sphere, and such like Things may remain, whatsoever happens without them, *i.e.* though all circumjacent Matter be any how changed, taken away, or annihilated. Which Things being lawfully supposed and laid down, the Reality of *Space* distinct from Magnitude does seem to be many Ways established.

As *first*, since Matter may be finite, and God is infinite in Essence, he must subsist beyond the Bounds of Matter, otherwise, he would be enclosed within its Limits or some Way bounded, and therefore could not be infinite. Therefore something is beyond, *i.e.* some sort of *Space*. And if God do not exist beyond the Bounds of Matter, our Imagination would be capable to conceive a Place where he is not; and consequently would in some Sort transcend the Manner of the Divine Essence; whence it would follow, that we could not apprehend or conceive God to be immense.

Next, God can create other Worlds beyond this, as also we can imagine them to be created, not no where, but some where; there will therefore be given some Space wherein they may be placed and exist. And God will also be present in these new produced Worlds, yet without being the least affected with Motion himself (for all kind of Immobility and Immutability is an undoubted Attribute of the Divine Perfection) which can no otherwise be understood, then by conceiving him to have before been present in the *Space*, where [**171**] they[24] are now reposited. Therefore

[23] See [p. 168] above.

[24] I.e., the new worlds.

the Conceptions which we have, or ought to have, concerning the Divine Infinity, Power, and Immutability do involve some distinct *Reality* of *Space*.

Moreover the material World, from the Hypothesis which we have asserted, will be limited and endowed with some Figure, and consequently may be supposed of any Figure. Let it therefore be Spherical; and because it may also be lawfully done, let some other spherical World, or Body, be in like Manner supposed to be contiguous to it, which will touch the former in only one Point; therefore there will lie some Mean, *i.e.* some *Space* between the other Points of the Spheres. For let any two Points be assumed in the Superfice of the contiguous Spheres without the Contact, I say some Space will lie between these; If you deny it, then these two Points will meet one another, contrary to a most clear Demonstration in Geometry.

Also let two Centers of the Spheres be connected by a Right Line passing through the Contact as Geometry also teaches and proves; and let two Radii be understood to proceed from the Centers to the said Points without the Contact; then because, according to the Opinion of our Adversaries, the said Points are contiguous, they will constitute a Triangle of three Right Lines, two of whose Sides will be equal to the third, in like manner contrary to a most clear and certain Theorem of Geometry. Again let the Matter, which goes to constitute two concentric Spheres, be supposed in any Place, and let the interjacent Matter between their Superficies be annihilated, or removed out of the Way, (which has been supposed may be done by the Power of God) then these Superficies, if no *Space* come between, will meet one another, though the one were a thousand times greater than the other: For we may suppose **[172]** from what has been said that both the spherical Superficies do retain their Magnitude, the Evacuation of the middle Body, which happens extrinsecally and changes nothing within them no ways hindering it; the Divine Power some how preserving their Magnitude and Position. In the same manner if what is between the Vertex and Basis of a Pyramid be taken away, since there is no Space left, therefore the Point of the Vertex will lie next to all the Points of the Basis, and consequently will be congruous and equal to the whole. These and innumerable such other Things do seem to follow, from the Denial of *real Space*, which are no less repugnant to the common Conceptions of Men, than [to] the Principles of Geometry.

Besides it seems somewhat harsh to suppose the Matter of the Universe entirely immoveable, as to the whole of it, or to imagine that nothing exists any where, except one solid Sphere; because such a Sphere could not indeed be transferred by God, not (to use the Words of *Plato*) *admit a Peripheron* or Rotation about its Axis,[25] which the taking away of Space does manifestly infer. For since in every such Motion, the Parts retain the same Situation to, and the same Distance from one another, howsoever the Whole be carried; it can be no otherwise conceived to be moved, but from a successive Change of *Space*, *viz.*so as one Part may enter the Space left by the foregoing one; from whence by the Denial of Space, we take away the Mobility of Matter. *Aristotle* in a certain Place observes something like this, where he says, *That Motion alone gives Occasion of Disquisition concerning Place*

[25] Possibly a reference to Plato's compressed description of the rotation of the 'Same' and 'Different'; see Lee (tr.), *Plato Timaeus*, pp. 49–50 and n.1.

or Space (as scarce being capable of Conception without the Position of Space) and the Heaven above all Things seems especially to be in Place, because it is most of all moved: nor yet is this any how varied (in respect of the first or principal diurnal Mo**[173]**tion, which the Philosopher seems to respect) according to the Situation and Distance of the Parts, but as the Whole is carried about. It is therefore so far moved, or not at all moved as its Parts do change their *Spaces*; and those Parts which are now at the Rising of the Sun, do anon, reach the height of the Meridian, and as soon as they have passed that do immediately decline to the Westward. Nor will it here suffice *Cartesius* to distinguish Action from Motion,[26] since no Effect of this Action can be imagined without a change of Space. ... **[174–175]** ...

Thus for the most part is the Matter disputed on both Sides; what then shall we finally determine? How shall we reconcile these contrary Probabilities and decline the Difficulties that on every Side beset us? I on my own Part will advance nothing to be true in a Case so nice and dubious, nor necessitated to give my Opinion, and compelled to declare what to me seems most to resemble the Truth, I would not be too averse to the common Conceptions of Men, nor oppose the sacred Laws of Geometry. I would say *first*, that *Space* is a thing really distinct from Magnitude; *i.e.* that something is designed by that Name, that a Conception answers it, that it is founded in the Nature of Things, that it is different from the Conception of Magnitude, and though Magnitude had no Existence at all, yet there would be Space. I would say, *secondly*, that Space is not **[p. 176]** any thing actually existent, and actually different from Quantity, much less that it has any Dimensions proper to itself, and actually separate from the Dimensions of Magnitude. You will perhaps say, What will it be then? What means this Riddle? I do not very much promise myself, nor dare hope that I shall satisfy you with mine Answer, but because Form requires it, I reply, that *Space* is nothing else but the *mere Power, Capacity, Ponibility*,[27] or (begging pardon for the Expressions) *Interponibility*[28] *of Magnitude*.[29] I thus explain my Meaning: Before the Creation of the World, there was no Body any where (as is reasonable and pious to believe) but yet it was possible for the greatest Body whatever then to exist, and obtain a determinate Position by the Will and Power of God, *i.e., there is Space*. There lies no Body, there is found no actual

[26] See Miller and Miller (trs.), *René Descartes Principles of Philosophy*, Pt.II §25, p. 51, regarding 'What movement is properly speaking', in which Descartes distinguished between 'a *transference*' and 'the force or action which transfers'.

[27] OED obs. rare: Capable of being placed (1734, Barrow the sole citation). See also Impone OED *v.* obs. 1. *trans*: To place or set upon something; to impose (1529–1729); 2. *intrans.*: To impose upon (1640).

[28] OED obs. rare: Capacity of being interposed (1734, Barrow the sole citation). See also Interpone OED *v.* obs. = Interpose *v.* I.1.b *trans.*: To place things with intervals or in alteration (1602–91).

[29] In his *Lectiones geometricæ*, published in 1670, Barrow also defined time (in its real or absolute sense) as a capacity thus: 'Time ... does not denote an actual existence but merely the capacity or possibility of a continuity of existence, just as Space denotes the capacity for intervening magnitude.' See Stone, *Isaac Barrow*, p. 136, who, pp. 135–8, reproduces from William Whewell's 1860 edition of the mathematical works, the whole of Barrow's lecture 'with but slight excisions'. See also Burtt, *The Metaphysical Foundations of Modern Science*, pp. 155–61.

Dimension beyond the Mass of the Universe; but it is possible for a Body to be constituted and a real Dimension to be extended beyond that itself, *i.e., there is an Ultramundane*[30] *Space*. If all the Matter be excluded by the divine Power from between these Walls, there will actually be no Body between them, but there will remain a Capacity of putting some Body between them, *i.e., there is an intermediate Space*. Lastly no Magnitude can be interposed between two adjacent or contiguous Magnitudes; *i.e. no Space or Interval is between them.*

Here it may be observed by the Way, that the Nature of every particular *Space* is in some sort *determinate*, and in some sort *indefinite*; *Determinate* as to the Mathematical Species and Quantity of its Figure; for it is not every Capacity whatsoever, but only the Capacity of a similar and equal Magnitude; *Indefinite* as to the other Qualities and **[177]** physical Species, also as to the Individuities of the Magnitudes, if I may so speak, for it is the Capacity of every Magnitude equally great and endowed with a similar Figure.... Hence every positive Interval endowed with actual Dimension and really extended of itself, divisible, terminated, pertransible,[31] or congruous to Bodies, is not denoted by the Word *Space*; but it only signifies that a Body may be so extended, may be so figured, is adaptable to such a Measure, and may exist either together in the same Instant or successively by Motion. I say it has no actual but only potential Figures, Dimensions and Parts consentaneous to its Nature; by which Means the Capacity of admitting a Body includes the Capacity of admitting Lines and Superficies, respectively.... Neither can the Quantity or Quality of *Space* be determined immediately or of itself, but by the Measure of Determination of some real Magnitude occupying it....

This Name therefore of Space is not of a mere Nothing or Thing feigned at pleasure, as of an *Hircocervus* or *Chimera,* but is deservedly to be placed in the same Order of Beings with *Creatibility, Sensibility, Mobility*, and such Possibilities; and there are scarce any but who adjudge some **[178]** sort of Reality to these. Nor do I see why this *Space* may not be a Being as well as *Contiguity*, to which it seems to be directly opposed. For *Contiguity* is the Mode of Magnitudes which signifies that no Magnitude can come between them but by moving them out of their Places: And on the contrary *Space* is that Mode of the same which intimates that some other Magnitude may be interposed without moving them out of their Place.

And this Notion of *Space* being supposed and granted, we may untie the Knots and remove the Difficulties of both the aforesaid Opinions.[32] For first nothing can be gathered hence derogatory to the prerogatives of the Divine Perfection, by supposing a Being [i.e., space] really eternal and infinite, unproduced and independent upon God; but rather his unlimited Power of producing and disposing Bodies at his Pleasure is asserted. Neither does the Idea of such a *Space* coincide with the Idea of

---

[30] OED *a.* B. Lying beyond or outside the world; of or belonging to things beyond the limits of the solar system.

[31] OED *a.* obs. rare: Capable of being traversed or passed through; traversible (1656).

[32] I.e., the opinions chiefly of Aristotle and Descartes. Note, however, that Barrow also included a critique of Hobbes's definition of space that is not included here; see Barrow, *The Usefulness of Mathematical Learning*, pp. 179–80.

Magnitude; but differs as far from it, as Power does from Act. Nor does it bring any other new real Beings into the Account besides Substance and Accident, but only denotes some Mode or Possibility of both. Nor will this *Space* require another Place; because it exists no where in an actual Manner, but will however be every where after its own Manner; because God can place Magnitudes any where. Nor is this contrary to the common Sense and Speech of Mankind, who, when they think or pronounce Space to intercede, do understand nothing but that some Body may be interposed between assigned Limits. Nor can any Infinity of Matter be deduced hence, but such Extension will follow as God shall please to assign it. Neither does this derogate from the Divine *Ubiquity*, which only signifies that God is present to all *Space*, or that something may exist every where. It also conspires and agrees with Geo[**179**]metry most precisely: for neither does this require that some actually real *Mean* shall always intercede between two *Points*, or any two *Terms*; but that sometimes, and in some Cases, a *Line*, *Superfice* or *Body* may intercede. It in like manner satisfies the Experiments and Phænomena of Natural Philosophers, affording them as much *Vacuum* as suffices for receiving Bodies, and performing their Motions; nor yet mingling any fictitious *Vacuum* endowed with real actual Dimensions such as *Epicurus* with his Followers dreamed to be one half of the Universe, and the first Principle of Bodies.[33] Again it appears hence that *Space* is immoveable and cannot be carried about with Bodies; because when one Body loses its other Confines or Interstices, yet this Possibility remains, and nothing hinders, but other Bodies equally near and intermediate may be substituted and succeed. ... [**180**] ...

But I know not what bewitching *Siren* has drawn me upon her Rocks, and held me entangled in her Nets, while I am doing what in me lies to avoid them. This out of the way Philosophy has put a stop to my Course, while I am failing with all my Speed to the principal Port of the Mathematics. That I may therefore finish this Digression concerning *Space* (indeed too prolix and spacious), and circumscribe it as a Figure within Bounds, I will only remark one Thing more conducing to my Purpose; *viz.* that whatsoever natural Philosophers do determine, this Method of conceiving *Space*, which I have been describing, is most agreeable, and abundantly sufficient for Geometricians:... **[181]** ...

To accommodate these Things to our Purpose, a Mathematician may frame such Hypotheses as these from what has been advanced. *That this or that Space may be filled*, i.e., *that Lines, Superficies or Bodies may be interposed between any assigned Points, Superficies or Bodies respectively, which, by Reason of some Property of their Nature, or precedaneous*[34] *Supposition, do not touch one another.* And on the

[33] According to Jones, *The Epicurean Tradition*, Epicurus imagined that 'body' and empty space are the sole existents of an infinite universe. But he distinguished between the terms 'universe' and 'world' (of which he imagined a plurality) by conceiving the former as unbounded and the latter as circumscribed within boundaries. Perhaps, therefore, Barrow's cryptic statement above refers to the bounded worlds, which are compounds generated from an infinite number of indivisible *minimæ partes* that constitute the permanent substratum of matter.

[34] OED *a.* obs.: Happening or existing before something else; preceeding, antecedent, previous (1647–1794).

contrary, *That every Magnitude filles Space*, i.e. *that a Magnitude cannot be placed between two Magnitudes, but the adjacent Magnitudes on either Side are separate and at a Distance, acccording to the Manner of the Magnitude supposed. That no Space is tied to any particular Magnitude but may be successively filled, as the Thing will suffer, by innumerable others, according to its Measure and the Exigence of the adjacent Magnitudes*; for, as before was explained,[35] *Space* is not a particular, but a **[182]** kind of general and indefinite Capacity: And reciprocally. *That no Magnitude is tied to any particular Space*. And which may be looked upon as a consequent Axiom from the foregoing Hypotheses, *That more Magnitudes than one taken together cannot adequately fill the same Space:* For more Acts than one do also argue more Powers than one, or one Act does perfectly fill up and exhaust only one Power; therefore more Magnitudes than one require more Spaces than one, or one Magnitude takes up the whole constitutive Capacity of only one Space.... And by what Reason any Plurality of Magnitudes do occupy one *Space*, by the same Reason may as many other Magnitudes as you please occupy the same Space; from whence the whole Infinity of possible Magnitude may be contained in the Space of one Grain of Barley; which seems to be inconsistent and contradictory to sound Reason. Also because Magnitudes may obtain the same Space, they will also possess the same Termination and Extension and other conjunct Affections of Magnitude, and consequently will become entirely the same; for we scarce do, or I believe, can conceive any Thing which constitutes or distinguishes Magnitude except these Affections

What this last Supposition or (as some chuse to call it) Axiom pronounces or signifies, is wont to be expressed by the Word *Impenetrability*, which many take the primary of all the Attributes of Magnitude, but how deservedly let them see themselves; for I am for no *Preference of Affections*, but do esteem all reciprocal and equal, as to *Degree*. Only it will be of use to us to take Notice, that this Position is very necessary to Mathemati**[183]**cians, and, as we have seen, is drawn from sound Reason, and is agreeable to perpetual Experience. For whatsoever in Mechanics is deduced from the *Pulsation*, or *pulsive Force*, of Bodies depends upon this; Also whatsoever in Geometry arises from the *Dependence of Motions* (of which kind is that most elegant Description of Curve Lines invented by *Cartesius* in his Geometry,[36]

[35] See [p.177] above.

[36] Descartes's, whose main goal in *De géométrie* was the application of algebra to geometry, applied this new method to the construction of curves in Bk. II; see Descartes, *Discourse on Method*, pp. 190–227. For a brief description of his method, see Mahoney in Crombie, Mahoney and Brown, 'René du Perron Descartes', *Dictionary of Scientific Biography*, vol. 4, p. 57. For Newton's response to this Cartesian method, see Guicciardini, *Isaac Newton*, p. 6, *et passim*. According to the notice of Barrow's *Lectiones geometricæ* in *Philosophical Transactions of the Royal Society* (1671), 6: 2260–2, five lectures were devoted to determining the '*Tangents* of *Curve* Lines, in the tenth of which he delivered 'a general Analytical method of determining Tangents, extending to all sorts of Curve lines, both *Geometrical* and *Mechanical* (as Monsieur *Des-Cartes* distinguisheth)'. If Barrow did deliver these lectures (and there is some dispute about this), then it is probable they were read after the last set of his Lucasian *Lectiones mathematicæ*; see Feingold, 'Isaac Barrow', pp. 68–9.

which is performed by the Impulse or Trusion of Rules in a certain ordinate Reason); because if a Plurality of magnitudes taken together can fill the same Space, or which is all one, can one penetrate another, then one would not necessarily give way to another, from whence the Supposition of Pulsation would be in vain, and no Effect would result, at least from a scientific Necessity. Therefore the *Impenetrability* of Magnitude is necessarily supposed by mathematicians, and consequently it was not foreign to the Purpose to shew its Agreement with Reason, and that it is not supposed unlawfully.

By equal Right it may on the contrary be supposed, that no particular Magnitude can occupy more Spaces than one taken together. ... And by what Reason one particular Magnitude can possess more *Spaces* than one, by the same Reason may it suffice to possess as many *Spaces* as you please....

I must not omit taking Notice, that the same Magnitude by occupying a Plurality of *Spaces* will obtain a Plurality of Extensions, Terminations, **[184]** and all other Affections of magnitude; from whence it will not continue one Magnitude, but become many. Neither is this Supposition unprofitable, but very necessary to Mathematicians: and whatsoever is supposed or demonstrated by them concerning Determinate Position depends on this. ... Whatsoever then may be said of the Popish [science of] Divinity,[37] except the Hypothesis be true, that it is not possible for the same Magnitude to fill more Spaces than one at the same Time, there is an entire End of all Geometry. But we cannot now add every thing worthy of Observation concerning *Determinate Position*; no more than concerning the other remaining Affections of Magnitudes which are to be reserved for the following Lectures.

I will only add this one Observation more, that there is a great Affinity and Analogy between *Space* and *Time*. For as *Space* is to *Magnitude*, so does *Time* seem to be to *Motion*; so that *Time* is in some sort the *Space* of *Motion*. ... **[185]** ... But this by the by, for *Time* itself now forbids us to treat more fully concerning *Time*.[38]

[37] For this cryptic remark, see Barrow, *A Treatise of the Pope's Supremacy*, p. 318: 'the greatest tyranny that ever was invented in the world' was 'the pretence of infallibility', for 'the Pope, not content to make us do and say what he pleaseth, will have us also to think so'; and such authority 'will inevitably produce depravation of Christian doctrine'. The doctrine of infallibility, which was formally defined in 1869–70 but defended earlier, was built on the Roman Catholic dogma of Petrine supremacy of the Pope; hence the title of Barrow's tract.

[38] For Barrow's lecture on time, see p. 147 n.29 above.

References

1. Primary Sources

Aristotle

see below Secondary Sources: Hett (tr.); Lawson-Tancred (tr.); Sorabji (tr.)

Boyle, Robert

The Works of the Honourable Robert Boyle ... ed. by Thomas Birch (new edn., 6 vols., London, 1772)

A Free Enquiry into the Vulgarly Received Notion of Nature ed. by E. B. Davis and Michael Hunter (Cambridge, 1996)

The Correspondence of Robert Boyle ed. by Michael Hunter, Antonio Clericuzio and L. M. Principe (6 vols., London 2001)

Barrow, Isaac

The Usefulness of Mathematical Learning explained and demonstrated: Being Mathematical Lectures read in the Publick Schools at the University of Cambridge ... tr. by John Kirkby (London, 1734)

A Treatise of the Pope's Supremacy [1680] ... ed. by Alexander Napier (Cambridge, 1859)

Briggs, William

Ophthalamo-graphia [1676]

Ophthalamo-graphia, sive Oculi ejusque partium descriptio (Cantabrigiæ, 1676)

Ophthalamo-graphia [1685]

Ophthalmo-graphia, sive Oculi ejusq; partium descriptio Anatomia. Cui accessit Nova Visionis Theoria, *Regiae Societati Londin. proposita.* Editio altera (London, 1685)

J. C. Kassler, *Newton's* Sensorium*: Anatomy of a Concept*, Archimedes 53,
https://doi.org/10.1007/978-3-319-72053-1

Ophthalamo-*graphia* [1685/1990][1]

Ophthalmo-graphia, or an Anatomical Description of the Eye and its Parts to which is added A New Theory of Vision ... tr. by Jeffrey Willis, ed. by D. M. Albert (Birmingham, Alabama, 1990)

'A New Theory' [1682]

'A New Theory of Vision presented to the Royal Society', *Philosophical Collections* (March 1681/2), no. 6, pp. 167–78

'A Continuation' [1683]

'A Continuation of a Discourse about Vision, with an Examination of some late Objections against it,' *Philosophical Transactions of the Royal Society* (10 May 1683), vol. 13, no. 147, pp. 171–82

Browne, Thomas

The Major Works ed. with an introduction and notes by C. A. Patrides (Harmondsworth, Middlesex, 1984)

Cicero

The Nature of the Gods tr. by H. C. P. McGregor with an introduction by J. M. Ross (Harmondsworth, Middlesex, 1984)

Clarke, Samuel

A Discourse concerning the Unchangeable Obligations of Natural Religion.... (London, 1706)

Cudworth, Ralph

The True Intellectual System of the Universe: The First Part; wherein All the Reason and Philosophy of Atheism is confuted; and its impossibility demonstrated (London, 1678)

Descartes, René

Discourse on Method, Optics, Geometry and Meteorology [French text 1637] tr. with an introduction by P. J. Olscamp (Indianapolis, Indiana, 1961)

Miller, V. R. and R. P. Miller (trs.)

René Descartes Principles of Philosophy [Latin text 1644, French text 1647] tr. with explanatory notes by V. R. Miller and R. P. Miller (Dordrecht, 1984)

Treatise on Man [Latin text 1662, French Text 1664] tr. by T. S. Hall (Cambridge, Mass., 1972)

'The Description of the Human Body and All its Functions, Those that do not depend on the Soul as well as Those that do. And also the principal Cause of the Formation of its Parts', *The World and other Writings* tr. and ed. by S. Gaukroger (Cambridge, 1998), pp. 170–205

[1] For some problems with this version, see *supra* Pt.III.3.1, p. 62 n.7 and p. 67 n.38.

Edwards, John
'Postscript', *Some Brief Critical Remarks on Dr. [Samuel] Clarke's Last Papers....* (London, 1714), pp. 36–40

Ficino, Marcilio
'Five Questions concerning the Mind', *The Renaissance Philosophy of Man, Selections in Translation* ed. by E. Cassirer, P. O. Kristeller, J. H. Randall, Jr. (Chicago, 1948), pp. 155–212

Galen
On the Doctrines of Hippocrates and Plato ed., tr. and comment by P. de Lacy, First part, Books I–V (2d edn., Berlin, 1981)

see also below Secondary Sources: Brock (tr.); May (tr.)

Gassendi, Pierre
Selected Works.... ed. and tr. by C.B. Brush (New York, 1972)

Halley, Edmund
'Of the Infinity of the Sphere of Fix'd Stars', *Philosophical Transactions of the Royal Society* (1720–21), 31: 22–24

Harvey, William
Lectures on the Whole of Anatomy, an annotated tr. of *Prelectiones Anatomiæ Universalis [1616–18]* by C. D. O'Malley, F. N. L. Poynter and K. F. Russell (Berkeley, 1961)

The Anatomical Exercises: De motu cordis[1628] *and De circulatione sanguinis* [1649] tr. by R. Lowndes [1653] and ed. by G. Keynes (New York, 1995)

Hobbes, Thomas
Leviathan [1651] ed. with an introduction by C. B. Macpherson (Harmondsworth, Middlesex, 1985)

Locke, John
An Essay concerning Human Understanding [4th edn, 1700] ed. by P. H. Nidditch (Oxford, 1985)

see also Secondary Sources: Dewhurst

Lower, Richard
Tractatus de corde [1669] ... tr. by K. J. Franklin [vol. 9 of R. T. Gunther, *Early Science in Oxford*], (Oxford, 1932)

Mariotte, Edme
'The Answer of Monsieur Mariotte to Monsieur Pecquet about the Opinion, that the Chorœides is the Principal Organ of Sight; Communicated to the Publisher from Paris, and Englished as Follows', *Philosophical Transactions of the Royal Society* (1670), 5: 1023–42

More, Henry

Philosophical Writings ... ed. by F. I Mackinnon (New York, 1969)

Newton, Isaac

'Of Colours' (c.1666), Cambridge University Library MS Add 3975: ff. 1–22; transcribed in J. E. McGuire and Martin Tamny, *Certain Philosophical Questions: Newton's Trinity Notebook* (Cambridge, 1983), Appendix, pp. 466–89

'De gravitatione et æquipondio fluidorum' *incipit* (product of lectures read in the early 1670s[2]), Cambridge University Library Add MS 4003: ff. 4–23v; transcription of the Latin with English tr. by A. Rupert Hall and Marie Boas Hall, *Unpublished Scientific Papers of Isaac Newton: A Selection from the Portsmouth Collection in the University Library, Cambridge* (Cambridge, 1962), pp. 89–156

'Conclusio' (c.1687), Cambridge University Library MS Add 4005: ff. 25–8, 30–7; transcription of the Latin with an English tr. by A. Rupert Hall and Marie Boas Hall, *Unpublished Scientific Papers of Isaac Newton: A Selection from the Portsmouth Collection in the University Library, Cambridge* (Cambridge, 1962), pp. 320–47

'Tempus et Locus' *incipit* (early 1690s?), Cambridge University Library Add MS 3965 section 13, ff. 541–541, 545–546; transcription of the Latin with English tr. by J. E. Mcguire, 'Newton on Place, Time and God: An Unpublished Source', *The British Journal for the History of Science* (1978), 11: 114–129, pp. 116–23

'De motu et sensatione Animalium' and 'De vita & morte vegetabili' (c.1709 or perhaps 1710–11),[3] Cambridge University Library MS Add 3970: ff. 236–240v; transcription and comment in Maurizio Mamiani and Emanuela Trucco, 'Newton e i fenomeni della vita', *Nuncius* (1991), 6: 69–96, pp. 77–9, 79–87

Opticks (1)

Opticks: or, A Treatise of the Reflexions, Refractions, Inflexions and Colours of Light.... (London, 1704)

Optice

Optice: sive de reflexionibus, refractionibus & coloribus lucis libri tres. Latine reddidit Samuel Clarke (London, 1706)

---

[2] For details relating to dating this manuscript, see Feingold, *The Newtonian Moment*, pp. 26–7.

[3] In place of the dates suggested by the transcribers and commentators, these dates are taken from Shapiro, 'Newton's Optical Theories and Vibrating Media', p. 36.

Opticks (2)
Opticks: or, A Treatise of the Reflections, Refractions, Inflections and Colours of Light. The Fourth Edition, corrected (London, 1730).

Opticks (C-R)
Opticks or, A Treatise of the Reflections, Refractions, Inflections & Colours of Light ...based on the fourth [1730] edition ... with a foreword by A. Einstein, an introduction by Sir E. Whittaker, a preface by I. B. Cohen, and an analytical table of contents prepared by D. H. D. Roller (New York, 1952)

Principia mathematica [1687]
Philosophiæ naturalis principia mathematica (London, 1687)

Principia mathematica (C-M)
Sir Isaac Newton's Mathematical Principles of Natural Philosophy and His System of the World tr. into English by Andrew Motte in 1729. The translations revised, and applied with an historical and explanatory appendix, by Florian Cajori (Berkeley, 1946)

The Principia
The Principia: Mathematical Principles of Natural Philosophy. The authoritative translation by I. Bernard Cohen and Anne Whitman assisted by Julia Budenz (Oakland, California, 1999)

The Correspondence
The Correspondence of Isaac Newton (7 vols., Cambridge, 1959–77)

see also below Secondary Sources: Cohen and Schofield (eds.); McGuire and Tamny (eds.)

Plato
see below Secondary Sources: Lee (tr.)

Willis, Thomas
Diatribæ duæ medico-philosophicæ [1659]
Diatribæ duæ medico-philosophicæ, quorum prior agit de fermentatione.... (London, 1659)

A Medical-Philosophical Discourse [1659 tr. 1681]
A Medical-Philosophical Discourse of Fermentation or, Of the Intestine Motion of Particles in every Body [with two other tracts, 'Of Fevers' and 'Of Urines']. Tr. into English by S[amuel] P[ordage] (London, 1681)

Cerebri anatome [1664][4]
Cerebri anatome: cui accessit nervorum descriptio et usus (London, 1664)

The Anatomy of the Brain [1664 tr. 1681][5]
'VI. Of the Anatomy of the Brain'. 'VII. Of the Description and uses of the Nerves', *The Remaining Medical Works ... With large Alphabetical Tables for the Whole, and an Index for Explaining all the hard and unusual Words and Terms of Art, derived from the Latine, Greek, or other Languages, for the benefit of the meer English Reader, and meanest capacity. With Eighteen Copper Plates* Englished by S[amuel]. P[ordage]. Esq (London, 1681), pp. 55–192

Pathologiæ cerebri, et nervosi [1667]
Pathologiæ cerebri, et nervosi generis specimen. In quo agitur de morbis convulsivis, et de scorbuto (Oxford, 1667)

An Essay of the Pathology of the Brain [1667 tr. 1684]
An Essay of the Pathology of the Brain and Nervous Stock: In which Convulsive Diseases are treated of. ... Translated out of Latine into English, by S. P[ordage] (London, 1684)

Affectionum que dicuntur ... pathologia spasmodica vindicata [1670]
Affectionum quæ dicuntur hystericæ et hypochondriacæ pathologia spasmodica vindicata, contra responsionem episteolarem Nathanael Highmori, cui accesserunt exercitationes medico-physicæ duæ. 1. De sanguinis accensione. 2. De motu musculari (London, 1670)[6]

De anima brutorum [1672]
De anima brutorum, quæ hominis vitalis ac sensitiva est, exercitationis duæ (London, 1672)

Two Discourses [1672 tr. 1683]
Two Discourses concerning the Soul of Brutes, which is that of the Vital and Sensitive of Man.... Englished by S. Pordage (London, 1683)

see also below Secondary Sources: Dewhurst

---

[4] A 272-page emended edition in 12mo (not the original 4to) was published in 1683 at Amsterdam by H. Wetstenius.

[5] Facsimile reprint in vol. 2 of William Feindel (ed.), *The Anatomy of the Brain and Nerves....* (2 vols, Montreal, 1965), re-issued, *The Anatomy of the Brain and Nerves....* (Birmingham, Alabama, 1978).

[6] For the translation of the two tracts, see Willis, 'IV. Of The Accension of the Blood. V. Of Musculary Motion', *The Remaining Medical Works* ... Englished by S. P. Esq [Samuel Pordage], (London, 1683), pp. 24–33, 34–39.

2. Secondary Sources

Alexander, H. G.
The Leibniz-Clarke Correspondence ... ed. with an introduction and notes (Manchester, 1956)

Algra, Kiempe
Concepts of Space in Greek Thought (Leiden, 1995)

Aronson, J. L., Rom Harré and E. C. Way
Realism Rescued: How Scientific Progress is Possible (London, 1994)

Bastholm, E.
The History of Muscle Physiology ... A Study of the History of Medicine (København, 1950)

'Niels Stensen's Myology', *Steno and Brain Research in the Seventeenth Century* ed. by Gustav Scherz (Oxford, 1968), pp. 147–53

Bennett, J. A.
'A Note on Theories of Respiration and Muscular Action in England *c.*1660', *Medical History* (1976), 20: 59–69

Bochner, Salomon
The Role of Mathematics in the Rise of Science (Princeton, New Jersey, 1981)

Boring, Edwin G.
A History of Experimental Psychology (2d edn., New York, 1929)

Sensation and Perception in the History of Experimental Psychology (New York, 1942)

Bowers, Brian
Sir Charles Wheatstone FRS 1802–1875 (London, 2001)

Brazier, Mary A. B.
'The Historical Development of Neurophysiology', *Handbook of Physiology: Section I: Neurophysiology* ed. by J. Field, H. W. Magoun, and V. I. Hall (Washington, D.C., 1959), vol. 1, pp. 1–58

Brock, John Arthur (tr.)
Galen on the Natural Faculties [incomplete tr.] (Cambridge, Massachusetts, 1916)

Brook, John
'Beyond Everything', *Times Literary Supplement* (20 May 2005), p. 84

Brown, Theodore M.
'Richard Lower', *Dictionary of Scientific Biography* ed. by C. C. Gillespie (14 vols., New York, 1970–76), vol. 8, pp. 523–7

Buchdahl, Gerd
Metaphysics and the Philosophy of Science: The Classical Origins: Descartes to Kant (Cambridge, Massachusetts, 1969)

Buchwald, Jed. A and Mordechai Feingold
Newton and the Origin of Civilization (Princeton, New Jersey, 2013)

Burke, Vernon J.
Will in Western Thought: An Historico-Critical Survey (New York, 1964)

Burtt, E. A.
The Metaphysical Foundations of Modern Physical Science (1924, 2/rev. 1931, Garden City, New York, 1954)

Cassini, Paolo
'Newton: The Classical Scholia', *History of Science* (1984), 22: 1–58

Cassirer, Ernst
The Problem of Knowledge: Philosophy, Science, and History since Hegel tr. by W. H. Woglom, rev. by C. W. Hendel (New Haven, 1969)

The Individual and the Cosmos in Renaissance Philosophy tr. by M. Domandi (New York, 1963)

Clarke, E. S.
'Brain Anatomy before Steno', *Steno and Brain Research in the Seventeenth Century* ed. by G. Scherz (Oxford, 1968), pp. 27–34

Clarke, Edwin
'The Doctrine of the Hollow Nerve in the Seventeenth and Eighteenth Centuries', *Medicine, Science and Culture* ... ed. by L. G. Stevenson and R. P. Multhauf (Baltimore, Maryland, 1968), pp. 123–41

Clarke, Edwin and L. S. Jacyna
Nineteenth-Century Origins of Neuroscientific Concepts (Berkeley, California, 1987)

Cohen, I. B. and R. E. Schofield (eds.)
Isaac Newton's Papers & Letters on Natural Philosophy and related Documents (Cambridge, Massachusetts, 1958)

Colie, Rosalie L.
Paradoxia Epidemica: The Renaissance Tradition of Paradox (Hamdon, Connecticut, 1976)

Collins, James
God in Modern Philosophy (Chicago, 1959)

Connolly, Patrick J.
'Newton and God's Sensorium', *Intellectual History Review* (2014), 24: 185–201

Crombie, A. C.
Science, Optics and Music in Medieval and Early Modern Thought (London, 1990)

Medieval and Early Modern Science (2 vols, revised edn., Garden City, New York, 1959)

Robert Grosseteste and the Origins of Experimental Science 1100–1700 (Oxford, 1953)

Crombie, A. C., M. S. Mahoney, and T. M. Brown
'René de Perron Descartes', *Dictionary of Scientific Biography* ed. by C. C. Gillespie (14 vols., New York, 1970–6), vol. 4, pp. 51–65

Crowther, Kathleen M. and Peter Barker
'Training the Intelligent Eye: Understanding Illustrations in Early Modern Astronomy Texts', *Isis* (2013), 104: 429–70

Davis, Audrey B.
Circulation Physiology and Medical Chemistry in England 1650–1680 (Lawrence, Kansas, 1974)

Dempsey, Liam
'Written in the Flesh: Isaac Newton on the Mind-Body Relation', *Studies in History and Philosophy of Science* (2006), 37: 420–41

De Smet, Rudolf and Karin Verelst
'Newton's Scholium Generale: The Platonic and Stoic Legacy—Philo, Justus Lipsius and the Cambridge Platonists', *History of Science* (2001), 39: 1–30

Dewhurst, Kenneth
Thomas Willis's Oxford Lectures (Oxford, 1980)

Willis's Oxford Case Book (Oxford, 1981)

'Some Letters of Dr. Thomas Willis (1621–1675)', *Medical History* (1972), 16: 63–6

John Locke (1632–1704) Physician and Philosopher: A Medical Biography with an Edition of the Medical Notes in his Journals (London, 1963)

Dick, Steven J.

Plurality of Worlds: The Origins of the Extraterrestrial Life Debate from Democritus to Kant (Cambridge, 1984)

Dobbs, B. J. T.

The Foundations of Newton's Alchemy: The Hunting of the Green Lyon (Cambridge, 1975)

'Newton and Stoicism', *The Southern Journal of Philosophy* (1985), supplement 23: 109–23

Dockrill, D. W.

'Spiritual Knowledge and the Problem of Enthusiasm in Seventeenth Century England', *The Concept of Spirit* ed. by D. W. Dockrill and R. G. Tanner (Auckland, New Zealand, 1985)

'The Heritage of Patristic Platonism in the Seventeenth Century English Philosophical Theology', *The Cambridge Platonists in Philosophical Context* ed. by G. A. J. Rogers et al (Dordrecht, 1997), pp. 55–77

Ducheyne, Steffen

'The General Scholium: Some Notes on Newton's published and unpublished Endeavours', *Lias: Sources and Documents relating to the Early Modern History of Ideas* (2006), 33: 223–74

Feingold, Mordechai

The Newtonian Moment: Isaac Newton and the Making of Modern Culture (New York, 2004)

'Isaac Barrow: Divine, Scholar, Mathematician', *Before Newton: The Life and Times of Isaac Barrow* ed. by M. Feingold (Cambridge, 1990), pp. 1–104

'Isaac Barrow's Library', *Before Newton: The Life and Times of Isaac Barrow* ed. by M. Feingold (Cambridge, 1990), pp. 333–72

'Newton, Leibniz, and Barrow Too: An Attempt at a Reinterpretation', *Isis* (1993), 84: 310–38

see also Buchwald, Jed A. *above*

Ferguson, John

Utopias of the Classical World (London, 1975)

Frank, Robert J., Jr.

'Thomas Willis', *Dictionary of Scientific Biography* ed. by C. C. Gillespie (14 vols., New York, 1970–6), vol. 14, pp. 404–9

Harvey and the Oxford Physiologists: Scientific Ideas and Social Interaction (Berkeley, 1980)

Frangenberg, Thomas
'*Auditus visu prestantior*: Comparisons of Hearing and Vision in Charles de Bovelles's *Liber de sensibus*', *The Second Sense: Studies in Hearing and Musical Judgment from Antiquity to the Seventeenth Century* ed. by C. Burnett, M. Fend and P. Gouk (London, 1991), pp. 71–94

Fulton, J. F.
The History of the Physiology of the Muscle (Baltimore, 1930)

Gabbey, Alan
'Henry More and the Limits of Mechanism', *Henry More (1614–1687): Tercentenary Studies* ed. by Sarah Hutton, with a biography and bibliography by Robert Crocker (Dordrecht, Holland, 1990), pp. 19–35

Gingerich, Owen
'Johannes Kepler', *Dictionary of Scientific Biography* ed. C. C. Gillespie (14 vols., New York, 1970–76), vol. 7, pp. 289–312

Grant, Edward
Much Ado about Nothing: Theories of Space and Vacuum from the Middle Ages to the Scientific Revolution (Cambridge, 1981)

Guicciardini, Niccolò
Isaac Newton on Mathematical Certainty and Method (Cambridge, Massachusetts, 2011)

'Digitizing Isaac Newton', *Isis* (2014), 105: 403–9

Hall, A. Rupert
'Henry More and the Scientific Revolution', *Newton, his Friends and his Foes* (Aldershot, 1993), [XIII] pp. 37–54

Hall, Thomas S.
'On Biological Analogs of Newtonian Paradigms', *Philosophy of Science* (1968), 35: 6–27

Hamlyn, D. W.
Sensation and Perception: A History of the Philosophy of Perception (London, 1961)

Hamou, Philippe
'Vision, Color, and Method in Newton's *Opticks*', *Newton and Empiricism* ed. by Z. Biener and E. Schliesser (Oxford, 2014), pp. 66–93

Harrison, John
The Library of Isaac Newton (Cambridge, 1978)

Harrison, Peter

'Was Newton a Voluntarist?', *Newton and Newtonianism: New Studies* ed. by J. E. Force and Sarah Hutton (Dordrecht, 1990), pp. 39–63

Newtonian Science, Miracles, and the Laws of Nature', *Journal of the History of Ideas* (1995), 531–53

'Voluntarism and Early Modern Science', *History of Science* (2002), 40: 63–89

'Voluntarism and the Origins of Modern Science: A Reply to John Henry', *History of Science* (2009), 47: 223–31

Henry, John

'Voluntarist Theology at the Origins of Modern Science: A Response to Peter Harrison', *History of Science* (2009), 47: 79–113

'Galen's Mistakes', *Times Literary Supplement* (27 November 2015), p. 33.

Henry, John and J. E. McGuire

'Voluntarism and Panentheism: The Sensorium of God and Isaac Newton's Theology', *The Seventeenth Century* (2017), pp. 1–25

Hesse, Mary

Forces and Fields: The Concept of Action at a Distance in the History of Physics (London, 1961)

Hett, W. S. (tr.)

Aristotle On the Soul, Parva Naturalia, On Breath (Cambridge, Massachusetts, 1957)

Hierons, Raymond and Alfred Meyer

'Willis's Place in the History of Muscle Physiology', *Proceedings of the Royal Society of Medicine* (1964), 57: 687–92

Holton, Gerald

'Presupposition in the Construction of Theories', *The* [University of Texas] *Graduate Journal* (1965), 7: 87–109

Hundert, E. M.

Philosophy, Psychiatry and Neuroscience. Three Approaches to Mind: A Synthetic Analysis of the Varieties of Human Experience (Oxford, 1990)

Jacob, Alexander

'The Neoplatonic Conception of Nature in More, Cudworth and Berkeley', *The Uses of Antiquity: The Scientific Revolution and the*

Classical Tradition ed. by S. Gaukroger (Dordrecht, 1991), pp. 101–21

Jammer, Max
Concepts of Space: The History of Theories of Space in Physics (Cambridge, Massachusetts, 1954)

Janiak, Andrew
Newton as Philosopher (Cambridge, 2008)

Jevons, William Stanley
The Principles of Science: A Treatise on Logic and Scientific Method [1874, 2/1877] with a new introduction by E. Nagel (New York, 1958)

Jones, Howard
The Epicurean Tradition (London, 1992)

Kassler, Jamie C.
'Man—A Musical Instrument: Models of the Brain and Mental Functioning before the Computer', *History of Science* (1984), 22: 59–92

Inner Music: Hobbes, Hooke, and North on Internal Character (London, 1995)

Music, Science, Philosophy: Models in the Universe of Thought (Aldershot, 2001)

'Musicology and the Problems of Sonic Abuse', *Music, Sensation, and Sensuality* ed. by L. P. Austern (New York, 2002), pp. 321–33

The Beginnings of the Modern Philosophy of Music in England: Francis North's *A Philosophical Essay of Musick* (1677) with comments of Isaac Newton, Roger North and in the *Philosophical Transactions* (Aldershot, 2004)

Seeking Truth: Roger North's Notes on Newton and Correspondence with Samuel Clarke c.1704–1713 (Farnham, Surrey, 2014)

Kirk, G. S., J. E. Raven and M. Schofield
The Presocratic Philosophers with a Selection of Texts (2d edn., Cambridge, 1983)

Koelbing, H. M.
'Ocular Physiology in the Seventeenth Century and its Acceptance by the Medical Profession', *Steno and Brain Research in the Seventeenth Century* ed. by Gustave Scherz (Oxford, 1968), pp. 219–24

Koyré, Alexandre
'Les Queries de l'Optique', *Archives Internationales d'Histoire des Sciences* (1960), 13: 15–29

Koyré, Alexandre and I. Bernard Cohen
'The Case of the Missing *Tanquam*: Leibniz, Newton & Clarke', *Isis* (1961), 52: 555–66

Kristeller, Paul Oscar
Renaissance Thought II: Papers on Humanism and the Arts (New York, 1965)

Lawson-Tancred, Hugh (tr.)
Aristotle De Anima (On the Soul) tr. with an introduction and notes (Harmondsworth, Middlesex, 1986)

Lee, Desmond (tr.)
Plato Timaeus and Critias tr. with an introduction and an appendix on *Atlantis* (Harmondsworth, Middlesex, 1971)

Plato The Republic tr. with an introduction (2d edn, revised, Harmondsworth, Middlesex, 1974)

Leyden, Wolfgang von
Seventeenth-Century Metaphysics: An Examination of Some Main Concepts and Theories (London, 1968)

Lloyd, G. E. R.
Polarity and Analogy: Two Types of Argumentation in Early Greek Thought (Bristol, 1987)

Lysaght, D. J.
'Hooke's Theory of Combustion', *Ambix* (1937), 1: 93–108

McAdoo, Henry R.
The Spirit of Anglicanism: A Survey of Anglican Theological Method in the Seventeenth Century (New York, 1965)

McGuire, J. E. and P. M. Rattansi
'Newton and the "Pipes of Pan"', *Notes and Records of the Royal Society of London* (1966), 21: 108–47

McGuire, J. E. and Martin Tamny (eds.)
Certain Philosophical Questions: Newton's Trinity Notebooks (Cambridge, 1983)

see also Henry, John *above*

Magoun, H. M.
'Development of Ideas relating the Mind with the Brain', *The Historical Development of Physiological Thought* ed. by C. McC. Brooks and P. F. Cranefield (New York, 1959), pp. 19–28

Malet, Antoni
'Barrow, Wallis, and the Remaking of Seventeenth Century Indivisibles', *Centaurus* (1997), 39: 67–92

Mandelbaum, Maurice
Philosophy, Science and Sense Perception: Historical and Critical Studies (Baltimore, 1966)

Margeau, Henry
The Nature of Physical Reality: A Philosophy of Modern Physics (New York, 1950)

May, Margaret Tallmadge (tr.)
Galen on the Usefulness of the Parts of the Body ... [a complete] tr. from the Greek with an introduction and commentary.... (2 vols., Ithaca, New York, 1968)

Meyer, A. and R. Hierons
'A Note on Thomas Willis' Views on the Corpus Striatum and the Internal Capsule', *Journal of the Neurological Sciences* (1964), 1: 547–54

Morris, Charles W.
Six Theories of Mind (Chicago, 1932)

Needham, Dorothy M.
Machina Carnis: The Biochemistry of Muscular Contraction in its Historical Development (Cambridge, 1971)

Osler, Margaret J.
'Baptizing Epicurean Atomism: Pierre Gassendi on the Immortality of the Soul', *Religion, Science, and Worldview: Essays in Honor of Richard S. Westfall* ed. by M. J. Osler and P. L. Farber (Cambridge, 1985), pp. 163–83

Osmond, Percy H.
Isaac Barrow: His Life and Times (London, 1944)

Pagel, Walter
'Medieval and Renaissance Contributions to Knowledge of the Brain and its Functions', *The History and Philosophy of Knowledge of the Brain and its Functions* ed. by F. N. L. Poynter (Oxford, 1958), pp. 95–114

Paley, William
Natural Theology, or Evidences of the Existence and Attributes of the Deity, collected from the Appearances of Nature (10th edn., London, 1805)

Priestley, F. E. L.

'The Leibniz-Clarke Controversy', *The Methodological Heritage of Newton* ed. R. E. Butts and J. W. Davis, (Oxford, 1970), pp. 35–56

Ray, Christopher

Time, Space and Philosophy (London, 1991)

Riba, Neil M.

'Cartesian Optics and the Mastery of Nature', *Isis* (1997), 88: 42–61

Robinson, F. G.

The Shape of Things Known: Sidney's Apology in its Philosophical Tradition (Cambridge, Massachusetts, 1972)

Ryle, Gilbert

The Concept of Mind (New York, 1965)

Schlick, Moritz

Space and Time in Contemporary Physics: An Introduction to the Theory of Relativity and Gravitation tr. by H. I. Brose with an introduction by F. A. Lindemann (New York, 1920)

Schneewind, J. B.

The Invention of Autonomy: A History of Modern Moral Philosophy (Cambridge, 1998)

Shapiro, Alan E.

'Kinematic Optics: A Study of the Wave Theory of Light in the Seventeenth Century', *Archive for History of Exact Sciences* (1973), 11: 134–266

'Newton's Definition of a Light Ray and the Diffusion Theories of Chromatic Dispersion', *Isis* (1975), 66: 194–210

'The Spectre of Newton's "Spectrum"', *From Ancient Omens to Statistical Mechanics....* ed. J. L. Berggren and B. R. Goldstein, (Copenhagen, 1987), pp. 183–92

'The Optical Lectures and the Foundations of the Theory of Optical Imagery', *Before Newton: The Life and Times of Isaac Barrow* ed. by M. Feingold (Cambridge, 1990), pp. 105–78

Fits, Passions, and Paroxysms: Physics, Method, and Chemistry and Newton's Theories of Colored Bodies and Fits of Easy Reflection (Cambridge, 1993)

'The Gradual Acceptance of Newton's Theory of Light and Color, 1672–1727', *Perspectives on Science* (1996), 4: 59–140

'Newton's "Experimental Philosophy"', *Early Science and Medicine* (2004), 9: 185–217

'Newton's Optical Theories and Vibrating Media: The Electrical Spirit of the General Scholium and the Electrical Queries of the *Opticks*', *Archives Internationales d'Histoire des Sciences* (2015), pp. 77–99

Simpson, A. D. C.
'Robert Hooke and Practical Optics: Technical Support at a Scientific Frontier', *Robert Hooke: New Studies* ed. by M. Hunter and Simon Schaffer (Woodbridge, Suffolk, 1989), pp. 38–61

Smith, Emilie Savage
'Galen's Account of the Cranial Nerves and the Autonomic Nervous System', *Clio Medica* (1971), 6: 77–98, 173–94

Sorabji, Richard (tr.)
Aristotle on Memory (London, 1972)

Spiegel, R. E.
Galen on Psychology, Psychopathology, and Function and Diseases of the Nervous System.... (Basel, 1973)

Galen on Sense Perception: His Doctrines, Observations and Experiments on Vision, Hearing, Smell, Taste, Touch and Pain, and their Historical Sources (Basel, 1973)

Stead, Christopher
Doctrine and Philosophy in Early Christianity: Arius, Athanasius, Augustine (Aldershot, 2000)

Stevenson, Lloyd G.
'Anatomical Reasoning in Physiological Thought', *The Historical Development of Physiological Thought* ed. by C. McC. Brooks and P. F. Cranefield (New York, 1959), pp. 27–38

Strong, E. W.
Procedures and Metaphysics: A Study in the Philosophy of Mathematical-Physical Science in the Sixteenth and Seventeenth Centuries (Berkeley, California, 1936)

'Newton and God', *Journal of the History of Ideas* (1952), 13: 147–67

Tamny, Martin
'Newton, Creation, and Perception, *Isis* (1979), 70: pp. 48–58

Temkin, Owsei
'Francis Glisson', *Dictionary of Scientific Biography* ed. by C. C. Gillespie (14 vols., New York, 1970–6), vol. 5, pp. 425–7

Toulmin, Stephen
'Criticism in the History of Science: Newton on Absolute Space, Time, and Motion', *The Philosophical Review* (1959), 68: 1–29, 203–27

Tracy, T. J.
Physiological Theory and the Doctrine of the Mean in Plato and Aristotle (Chicago, 1969)

Upton, G. P.
Women in Music (London, 1909)

Van Helden, Albert
Measuring the Universe: Cosmic Dimensions from Aristarchus to Halley (Chicago, 1985)

Wallace, Wes
'The Vibrating Nerve Impulse in Newton, Willis and Gassendi: First Steps in a mechanical Theory of Communication', *Brain and Cognition* (2003), 31: 66–94

Wallis, Peter and Ruth Wallis
Newton and Newtoniana 1672–1975: A Bibliography (Folkestone, 1971)

Webster, Charles
'The Discovery of Boyle's Law, and the Concept of the Elasticity of Air in the Seventeenth Century', *Archive for History of Exact Science* (1962–66), 2: 441–501

Westfall, Richard S.
Never at Rest: A Biography of Isaac Newton (20th printing, Cambridge, 2010)

Wheatstone, Charles
'Contributions to the Physiology of Vision.—Part the First. On some Remarkable and hitherto unobserved Phenomena of binocular Vision, *Philosophical Transactions of the Royal Society* (1838), 128: 371–91

Whewell, William
Novum Organon Renovatum. Being the Second Part of the Philosophy of the Inductive Sciences (3d edn, with large additions, London, 1858)

Wightman, W. P. D.
The Emergence of Scientific Medicine (Edinburgh, 1971)

Wiles, Maurice
Archetypal Heresy: Arianism Through the Ages (Oxford, 2004)

Woollam, D. H. M.
'Concepts of the Brain and its Functions in Classical Antiquity', *The History and Philosophy of Knowledge of the Brain and its Functions* ed. by F. N. L. Poynter (Oxford, 1958), pp. 5–18

Yolton, John
Thinking Matter: Materialism in Eighteenth-Century Britain (Oxford, 1983)

Zafiropulo, Jean et Catherine Monod
Sensorium Dei, dans l'hermétisme et la science (Paris, 1976)

Zaitsev, Evgeny A.
'The Meaning of Early Medieval Geometry: From Euclid and Surveyors' Manuals to Christian Philosophy', *Isis* (1999), 90: 522–53

Index of Names

J. C. Kassler, *Newton's* Sensorium*: Anatomy of a Concept*, Archimedes 53,
https://doi.org/10.1007/978-3-319-72053-1

N

O

P

Z

Index of Subjects

J. C. Kassler, *Newton's* Sensorium*: Anatomy of a Concept*, Archimedes 53,
https://doi.org/10.1007/978-3-319-72053-1

W

The manufacturer's authorised representative in the EU is Springer Nature Customer Service Centre GmbH, Europaplatz 3, 69115 Heidelberg, Germany. If you have any concerns regarding our products, please contact ProductSafety@springernature.com

Printed and bound by CPI Group (UK) Ltd, Croydon, CR0 4YY
05/10/2026
02228135-0004